Phil O'Brien

SHEET METAL DRAWING
AND PATTERN DEVELOPMENT

Sheet Metal Drawing and Pattern Development

A. DICKASON

C.G.I.A., A.C.T (Birm.), M.Inst.S.M.E.

PITMAN PUBLISHING

First published 1967
Reprinted 1974

SIR ISAAC PITMAN AND SONS LTD.
Pitman House, Parker Street, Kingsway, London, WC2B 5PB
P.O. Box 46038, Banda Street, Nairobi, Kenya

SIR ISAAC PITMAN (AUST.) PTY. LTD.
Pitman House, 158 Bouverie Street, Carlton, Victoria 3053, Australia

PITMAN PUBLISHING CORPORATION
6 East 43rd Street, New York, N.Y. 10017, U.S.A.

SIR ISAAC PITMAN (CANADA) LTD.
495 Wellington Street West, Toronto, 135, Canada

THE COPP CLARK PUBLISHING COMPANY
517 Wellington Street West, Toronto, 135, Canada

© A. Dickason 1967

ISBN 0 273 41163 2

Made in Great Britain at The Pitman Press, Bath
G4(T1129/76)

Preface

It is not intended that this book should vie with *Geometry of Sheet Metal Work* for recognition as the instructional work on pattern development; it should be regarded as supplementary to that volume. The present work is not planned in progressive courses, though it is essential that it should progress in systematic order from the simple to the more difficult. Therefore, first is a review of the three methods of development, the radial line, parallel line, and triangulation, as applied to appropriate examples. Later, in Chapter 8, the problems of intersections, typical of many which occur in actual practice, provide ample opportunity for more advanced study on auxiliary, double, or multiple projections.

The chapter on Draughtsmanship in Sheet Metal Work presents a method, which the author used for many years, of laying out patterns for the craftsman which considerably simplifies the transference from paper to metal. It can also be very effective in saving material in the placing of the patterns on the sheets. The method is essentially a draughtsman's job, as much as is the original design and laying out of sheet metal work.

Two or three of the problems presented in this book originally appeared in *Sheet Metal Industries* Journal and I am indebted to the Editor, Mr. Edward Lloyd, for permission to use them in this work. I am also grateful for the continued help and interest of my wife and daughter in the preparation of this book.

A. DICKASON

Contents

		PAGE
Preface		v
1	POLYGONS AND CURVES	1
2	PROJECTIONS	20
3	RADIAL LINE DEVELOPMENT	30
4	PARALLEL LINE DEVELOPMENT	65
5	THE COMMON CENTRAL SPHERE	97
6	DEVELOPMENT BY TRIANGULATION	109
7	INTERSECTIONS BY CUTTING PLANES	165
8	AUXILIARY AND DOUBLE PROJECTIONS	185
9	DEVELOPMENT OF COMPLEX PATTERNS AND SPIRAL CHUTES	254
10	DUCT LAYOUT WITH DEVELOPMENTS	286
11	DRAUGHTSMANSHIP IN SHEET METAL WORK	345
Index		363

1 Polygons and Curves

REGULAR POLYGONS

As a preliminary to pattern development it is often necessary to construct a polygon either inside or outside a circle. Most of them are simple enough, but the following examples range progressively from the triangle to the octagon.

To construct an *equilateral triangle* inside a given circle, first draw the circle, then, with the same radius, describe arcs from the extremity A of any diameter AB to cut the circle in C and D, as shown in Fig. 1. Join CD, CB and BD.

To construct a *square* inside a given circle, draw two diameters at right angles to each other, as at AB and CD, Fig. 2, then join the extremities by chords, which completes the square.

To construct a *pentagon* inside a given circle, divide the circle into four quadrants, as in Fig. 3. Bisect OA in B, and with radius BC describe an arc CD cutting the diameter in D. Then, with radius CD describe an arc DE cutting the circle in E. The chord CE represents one side of the pentagon, which may be completed by stepping four more sides round the circle as shown in the Figure.

To construct a *hexagon* inside a given circle, draw a diameter, as at AB, Fig. 4. Then with the radius of the circle in the compasses and using A as centre, describe arcs cutting the circle in 1 and 2. Similarly, using B as centre, describe arcs cutting the circle in 3 and 4. Chords drawn between the points around the circle, as shown in the Figure, complete the hexagon.

To construct a *heptagon* inside a given circle, draw two diameters at right angles to each other, as shown in Fig. 5. Divide one half of the horizontal diameter into four equal parts, and add one of those parts to the diameter produced, as from O to A, Fig. 5. Join A to B. Then, from centre O, draw OC parallel to AB. The chord BC then forms one side of the heptagon, which may be completed by stepping that distance around the circle. This method is not perfectly accurate, since the angle subtended by the chord at the centre of the circle is one-twelfth of a degree short, but the method is simple and is sufficiently accurate for most practical purposes.

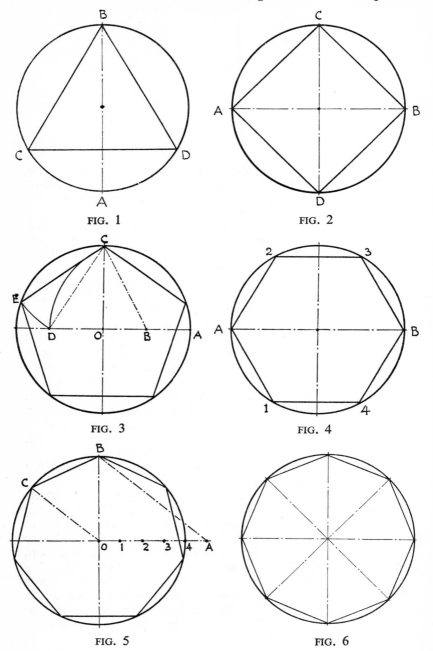

FIG. 1

FIG. 2

FIG. 3

FIG. 4

FIG. 5

FIG. 6

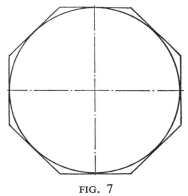

FIG. 7

To construct an *octagon* inside a given circle, draw four diameters at 45° to each other and join the extremities by a series of eight chords, as shown in Fig. 6.

Alternatively, an octagon may be readily drawn outside a circle by the use of a tee square and a 45° set square, as shown in Fig. 7. In this case the sides of the octagon are drawn as tangents to the circle, and no constructional details are needed when drawn in this way.

THE ELLIPSE

There are very many occasions associated with sheet metal pattern development when it is necessary to draw an ellipse. Numerous cross-sections of a cone present ellipses. A cylinder cut at any angle other than 90° to its central axis presents an ellipse. There are also numerous methods of drawing true ellipses, the choice of which depends largely on immediate requirements.

A true ellipse is a closed curve such that when two lines are drawn to the circumference from each of two points on the major axis, the two lines together are equal in length to the major axis. Each of the points on the major axis is called a focus, plural foci.

Referring to Fig. 8, the two points F' and F'' represent the foci. Then the distance from F' to 1, plus 1 to F'', is equal to the length of the major axis AB. Similarly, the distances F' to 2 and 2 to F'' are equal to AB. Again, F' to 3 plus 3 to F'' equals AB.

Since the positions of F' and F'' are equidistant from the centre O, and point 3 is at the extremity of the minor axis, it must be evident that the distances $F'3$, and $3,F''$ are equal. Also, since $F'3$ and $3,F''$ are together equal to the major axis AB, therefore $F'3$ must be equal to AO and $3,F''$ equal to OB. This furnishes a ready means of finding or locating the positions of F' and F''.

First draw the major and minor axes of the ellipse, Fig. 8. Then take half the major axis, AO, and from point 3 mark off the distances $3,F'$ and $3,F''$. The points F' and F'' then represent the foci of the ellipse. Next, divide the distance between F' and O into any number of parts, not necessarily equal, according to the number of points required to be located on the circumference of the ellipse. In Fig. 8, $F'O$ is divided into three parts by the points $1'$ and $2'$, though normally several more divisions would be advisable.

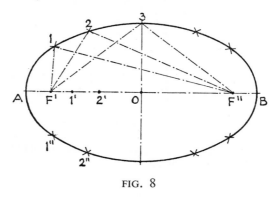

FIG. 8

Now, to find the corresponding points on the circumference, take the distance from A to $1'$, and using F' as centre describe arcs through points 1 and $1''$. Repeat this, with the same radius, from the focus F''. Next take the other part of the major axis from point $1'$ to B, and this time using F'' as centre describe arcs cutting the previous arcs in points 1 and $1''$. Then using the same radius and F' as centre, describe arcs to obtain similar points at the other end of the ellipse.

The whole process repeated, using the next two portions of the major axis from A to $2'$ and from $2'$ to B, will give the points 2 and $2''$ on the circumference, with a corresponding two points on the other side of the minor axis. By using more divisions between F' and O, a greater number of points may be obtained on the circumference, thereby enabling the curve to be drawn with greater accuracy.

No part of the curve of a true ellipse is a part of a circle, which means that a true ellipse cannot be drawn with the compasses. The curve should be drawn freehand through as many points as it is convenient to plot. French curves may be used, though these do not always result in an accurate ellipse unless the number of points

Polygons and Curves 5

obtained on the circumference is sufficient to ensure an adequate guide to the curve.

Approximate ellipses, or ovals, may be drawn with the compasses, though any curve or arc drawn with the compasses must necessarily be a part of a circle. Therefore, such constructions cannot be true ellipses. Nevertheless, some of them are very close approximations and a few useful methods are dealt with in *Geometry of Sheet Metal Work*.

There are a number of useful methods of plotting a true ellipse, one of which is shown in Fig. 9. Two circles are drawn from the

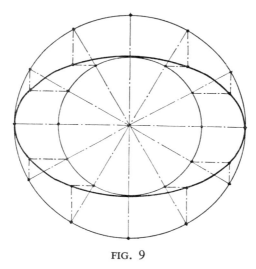

FIG. 9

same centre, the diameter of the smaller circle being equal to the minor axis of the ellipse, and that of the larger circle equal to the major axis of the ellipse. In the example given in the Figure the larger circle is divided into twelve equal parts, though the greater the number of divisions the more accurate the resulting ellipse should be, and it is not in fact necessary for the divisions to be equal. A series of diameters are now drawn from the points on the outer circle which cut through similar points on the inner circle. Next, vertical lines are drawn from the points on the outer circle to meet horizontal lines drawn from the corresponding points on the inner circle. The points where the verticals meet the horizontals lie on the circumference of the ellipse.

Another method of drawing a true ellipse is shown in Fig. 10. A rectangle is drawn with sides equal to the major and minor axes of the ellipse. The centrelines then represent the axes. The ends of the rectangle are divided into any equal number of equal parts, and the major axis is divided into the same number of equal parts, in this case eight. Then, from each end of the minor axis, radial lines are drawn as shown in Fig. 10. Where the set of lines drawn through

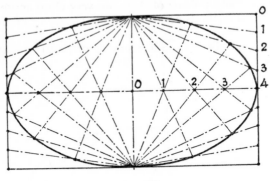

FIG. 10

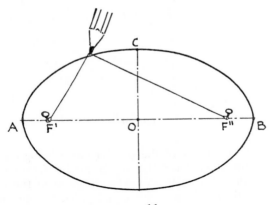

FIG. 11

the points on the major axis meet the set of lines drawn to the ends of the rectangle, points are located through which the circumference of the ellipse may be drawn. The construction should be readily followed from Fig. 10.

Fig. 11 shows a practical method of drawing a true ellipse. The method is based on the fundamental principle illustrated in Fig. 8, i.e. that the sum of the lines drawn from the foci to any point on the circumference is equal to the length of the major axis. In the

Polygons and Curves

present construction, the major and minor axes are drawn and the positions of the foci determined as explained in relation to Fig. 8. Two pins are then fixed on the points of the foci, and a piece of cotton, looped at each end to fit over the pins, is made to exactly the length of the major axis, loops included. The loops are slipped over the pins and a pencil point placed so that the cotton is stretched taut

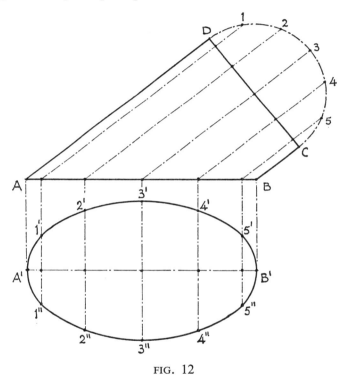

FIG. 12

as illustrated in Fig. 11. If the pencil is then moved while still keeping the cotton taut, an ellipse is traced out by the point.

In so far as sheet metal pattern development is concerned, the method of drawing a true ellipse shown in Fig. 12 is undoubtedly of the utmost importance as this construction occurs in many problems involving cross-sections or intersections. The Figure represents a cylinder of diameter CD, which is cut obliquely at AB. The section at AB is therefore an ellipse. This ellipse is shown plotted below with its major axis, $A'B'$, equal to AB, and its minor axis equal to the diameter CD of the cylinder. A semicircle is drawn on the diameter CD and divided into six equal parts. From the points on the semicircle, lines are projected back to CD, and on to the cut section at AB. From the points obtained on AB, lines are projected at right angles to AB for plotting the ellipse. Next, the major diameter $A'B'$ of the ellipse is drawn in any convenient position below and parallel to AB. The points on the circumference of the ellipse are now obtained from the semicircle on CD. The various widths from the diameter CD to the points 1, 2, 3, 4 and 5, are taken and marked off above and below the major axis $A'B'$ on the corresponding lines, as shown at $1'$, $2'$, $3'$, $4'$, $5'$, and $1''$, $2''$, $3''$, $4''$, $5''$. The curve of the ellipse may then be drawn through these points.

It will be evident that by dividing the semicircle on CD into a greater number of parts, and obtaining a correspondingly larger

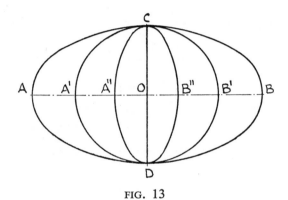

FIG. 13

number of points on the circumference of the ellipse, a more accurate curve may be plotted. Nevertheless, the practice of dividing a semicircle into six equal parts is a very convenient method, since the radius of the semicircle is used as a means of obtaining the six equal parts. A little care is needed in the freehand drawing of an elliptical curve in order to obtain a good shape without "flats" or "bulges."

In some respects a circle may be regarded as a special form of ellipse. Consider the diagram in Fig. 13, and imagine that the larger ellipse $ACDB$ on the major axis AB is rotated slowly about its minor axis CD. The major axis will get shorter while the minor axis remains unaltered. At one position $A'CB'D$ in the rotation, the horizontal axis will become equal to the vertical axis, or $A'B'$ equals CD. At this position the ellipse becomes a circle, and the foci fall on the diameter $A'B'$. If the larger ellipse continues to rotate, then from the position at which it becomes a circle, the horizontal axis becomes shorter than the vertical axis and the ellipse assumes the vertical position $A''CB''D$, CD now becoming the major axis. With continued rotation the horizontal axis vanishes when the points A and B reach the centre O. The ellipse then becomes a straight line corresponding to the axis CD.

THE PARABOLA

The parabola is a curve often encountered as a section of a cone in sheet metal pattern development. In simple terms it may be said

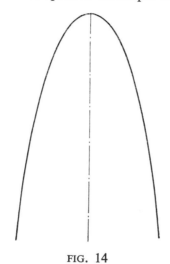

FIG. 14

that, starting at the vertex, a parabolic curve, curves round towards the centreline in such a way that the two sides approach closer and closer to the condition of being parallel to the centreline without ever quite becoming so, as may be seen in Fig. 14. A parabola may be defined as the form traced out by a point which moves in such a manner that its distance from a fixed point called the focus is equal to its distance from a straight line called the directrix.

Referring to Fig. 15, let the straight line AB represent the directrix, and F the focus. Then, by definition, any point on the curve is equidistant from the focus F and the directrix AB. For example, the distance $a'F$ is equal to $a'A$, and $b'F$ is equal to $b'b$. Also, $c'F$ equals $c'c$, $d'F$ equals $d'd$, and VF equals VD. It will be noted that the point *V* is the vertex of the curve, and is also the middle point between the focus *F* and the point *D* where the perpendicular from *F* cuts the directrix.

On the right-hand side of Fig. 15, it will be observed that the ordinates from the directrix *AB* (shown chain dotted) have been continued on the inside of the parabola as a series of full lines. These parallel lines could represent the reflected rays from a parabolic reflector when the source of light is placed at *F*, for rays from a light source placed at the focus of a parabolic reflector always emerge as a parallel beam.

There are numerous methods of drawing a parabola, two of which are shown in Figs. 16 and 17. In Fig. 16, let *AB* represent the

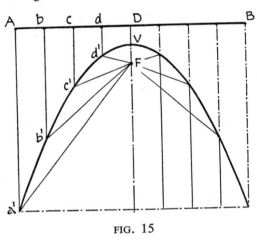

FIG. 15

directrix, and *D* be any point on *AB* from which *D*7 is drawn at right angles to *AB*. Let *F*, a point lying on *D*7, represent the focus of the parabolic curve. Then the vertex *V* is midway between *D* and *F*. Beginning anywhere below *V*, the distance *V*7 is divided into any number of parts, not necessarily equal, by lines drawn at right angles to *V*7. Next, the distances from *D* on the directrix are taken to each of the points thus marked on *V*7, and using *F* as centre, the distances are marked off on the corresponding cross lines. For example, *D*1 is taken, and, with *F* as centre, this distance is marked off on the cross line through point 1. Next, *D*2 is taken, and again using *F* as centre, the distance is marked off on the cross line through point 2. This process is repeated with each of the points down to 7, always using the focus point *F* as centre for marking off the distance through the corresponding cross line. The curve is

Polygons and Curves

then drawn through the points obtained on the cross lines as shown in Fig. 16.

The method shown in Fig. 17 is probably the most popular one for plotting, a parbolic curve. A rectangle is drawn corresponding

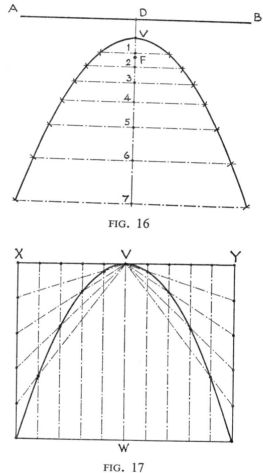

FIG. 16

FIG. 17

to the depth and width of the curve required. On each side of the centreline at V, each half of the width XY is divided into a number of equal parts. Each of the two sides representing the depth is divided into the same number of equal parts, not necessarily of the same size, as shown in the Figure. Lines are drawn from the width

XY parallel to the central axis VW. Then, from the point V, lines are drawn to the points down each side. The points where these oblique lines cross the corresponding parallel lines afford points through which to draw the parabolic curve. The construction should be readily followed from Fig. 17.

There is one important point to note in connexion with this construction. The oblique construction lines are drawn from the vertex, and not from the focus of the curve as in Fig. 15. Nevertheless, the resulting curve is correct, and the positions of the focus and the directrix may be found if required. The method of finding

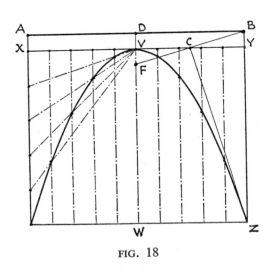

FIG. 18

these points is shown in Fig. 18. On the left-hand side of the Figure the construction is shown as in Fig. 17. On the right-hand side the construction is somewhat abbreviated in order to show the method of finding the position of the focus F and the directrix AB more clearly. First, the centre C of VY is located and joined to the bottom point Z of that side of the figure. Then a line is drawn through C at right angles to CZ. The line FCB cuts the axis VW of the parabola at the focus F and also cuts ZY extended at a point B which lies on the directrix. The directrix AB may now be drawn through B parallel to XY. It will be found, on checking, that each of the points on the curve conforms to the previous definition, that the distance from the point to the focus is the same as the distance from the point to the directrix.

Polygons and Curves

THE HYPERBOLA

The hyperbola is another curve frequently encountered as a section of a cone in sheet metal pattern development. This curve is somewhat similar in appearance to the parabola though it differs in its basic property. The hyperbola is related to two lines forming an angle. The curve, starting from a vertex within the angle, moves round towards the sides of the angle, always getting nearer to each side but never touching or coinciding with either, as may be seen in Fig. 19.

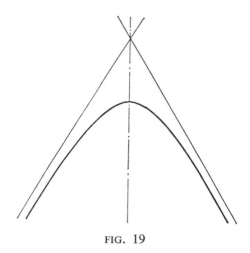

FIG. 19

This property may perhaps be best illustrated by reference to a section of a cone. When a cone is cut by a plane which is parallel to, but not through, its central axis, as at OR, Fig. 20, the true shape of the section is a hyperbola. In the plan, the cutting plane is a straight line across the circular base at a distance D from the centre. In the side elevation the shape of the hyperbolic curve made by the cutting plane is shown with its vertex at O, and its width A at the base. The width A is somewhat smaller than the diameter B of the base. These dimensions are shown also as A and B in the plan.

Imagine now that the cone be extended or produced downwards indefinitely, with the cutting plane still remaining at the distance D from the central axis of the cone. The base diameter B will obviously increase indefinitely, as also will the width A across the base of the hyperbola. Observe, now, that as the diameter increases, the width

14 *Sheet Metal Drawing and Pattern Development*

A must approach more and more to the diameter, but can never actually become equal to it.

Mathematically, the hyperbola has a focus and a directrix, somewhat similar to those of the parabola, but in the case of the

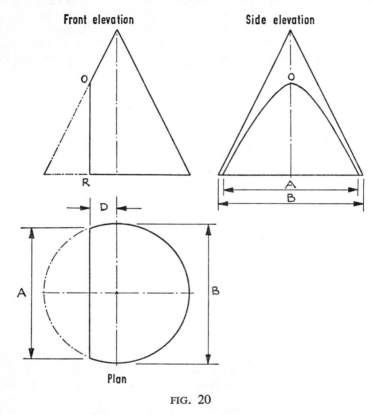

FIG. 20

hyperbola the vertex of the curve does not lie midway between the focus and the directrix.

CONIC SECTIONS

Fig. 21 shows a right cone with its circular base at *AB*. Let a line perpendicular to the plane of the paper at *O* be the centre of rotation of a cutting plane in which the first position at *OP* is perpendicular to the side *AC*; the second position *OQ* is parallel to the side *BC*; and the third position *OR* is perpendicular to the base *AB*. These

Polygons and Curves

three sections present an ellipse at *OP*, a parabola at *OQ*, and a hyperbola at *OR*. The methods of projecting the true shapes of these cross-sections are somewhat similar, and are shown separately in Figs. 22, 23 and 24.

Referring to Fig. 22, a semicircle is drawn on the base *AB*, which represents a half plan of the cone. The semicircle is divided into six equal parts, at 1, 2, 3, 4, and 5, and the points projected perpendicularly back to the base *AB*. From the points on the base, lines are drawn to the apex *C*. These represent conic lines on the surface

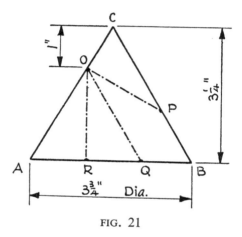

FIG. 21

of the cone. Corresponding conic lines are now drawn in the plan from the points 1, 2, 3, 4 and 5 on the semicircle to the plan apex *C'*.

The next step is to obtain the half plan of the ellipse by dropping lines vertically from the points on *OP* to meet the corresponding radial lines in the plan. Thus, *O* and *P* are obtained as *O'* and *P'*, and the remaining points occur on the radial lines from 1, 2, 4 and 5. It will be seen that the point on the centreline 3,*C'* cannot be obtained by dropping a vertical line from the point on *OP*, which is also on the centreline. Therefore a horizontal line from that point is drawn to the outside of the cone, and a line dropped from the point on the outside of the cone to the base *AB*. Then from the centre *C'*, the point obtained on the base *AB* is swung round to the line 3,*C'*, thus completing the required number of points on the radial lines. The elliptical curve from *O'*, through the points on the radial lines to *P'*, then represents the half plan of the ellipse.

Now, to obtain the true shape of the ellipse, project lines at right angles to *OP* from all the points on *OP*. In any convenient position draw the centreline $O''P''$ across and at right angles to the projected lines. The next step should be carefully observed. The width of the ellipse on either side of the centreline across the section at OP is the same as that in the half plan from the line AB. Therefore the perpendicular distances from AB to the points on the ellipse in the half plan are taken and marked off on the corresponding projected lines on both sides of the centreline $O''P''$. Consider, for example,

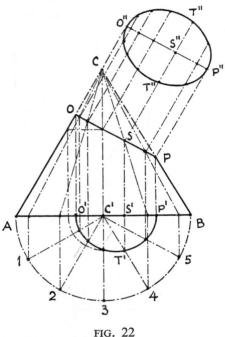

FIG. 22

the point S on OP. The width of the ellipse at that point is represented by $S'T'$ in the plan. Therefore the distance $S'T'$ is taken and marked off on both sides of the centre-line $O''P''$ as at $S''T''$ in the projection. This process is repeated with each of the other points on the ellipse, thus providing sufficient points in the projection through which to draw the true shape of the ellipse.

Referring to Fig. 23, the line OQ represents the position of the parabolic section. The true shape of the parabola is projected at right angles to OQ, as will be seen at $Q''O''Q''$. The first step is to describe a semicircle on the base AB. The semicircle again represents a half plan of the cone, and is divided into the usual six equal parts,

Polygons and Curves

although in this case only the quadrant from *A* to 3 will be needed. Points 1, 2 and 3 are projected perpendicularly back to the base *AB*, and from the points obtained on the base, lines are drawn to the apex *C*. These represent radial lines on the surface of the cone, and the corresponding plan lines are drawn from points 1, 2 and 3 to the apex *C'* in the plan.

The half plan of the parabola is now obtained by dropping lines verically from the points where the radial lines in the elevation

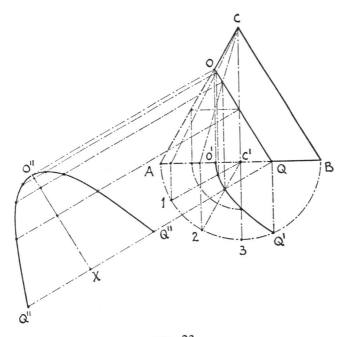

FIG. 23

cross *OQ* to meet the corresponding radial lines in the plan. First, *O* and *Q* are obtained as *O'* and *Q'*, followed by two further points on the lines 1,*C'* and 2,*C'*. As in the previous example, it will be seen that the required point on the centreline 3,*C'* cannot be obtained by dropping a vertical line from the point on *OQ* which also occurs on the centreline. Therefore the point on *OQ* is projected to the side of the cone, and vertically downwards to the base *AB*, and from *C'* the point is swung round to the centreline 3,*C'*. The half plan of the parabola may then be drawn from *O'* to *Q'* as shown in Fig. 23.

Now, to obtain the true shape of the parabola, lines are projected at right angles to OQ from all the points on OQ, and a centreline drawn across the projected lines in any convenient position as may be seen at $O''X$, Fig. 23. Next, the various widths of the curve are obtained from the half plan, by taking the perpendicular distances from AB to the points on the curve and marking them off on both sides of $O''X$ on the corresponding lines in the projection. Thus, sufficient points are afforded through which to draw the parabolic curve as shown in Fig. 23.

Referring now to Fig. 24, the line OR represents the position of the hyperbolic section, and again the true shape of the curve is

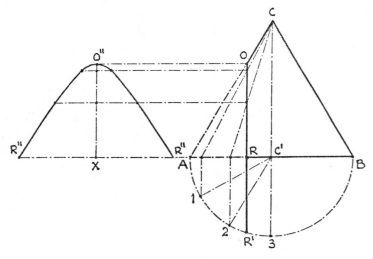

FIG. 24

projected at right angles to the position of the section in the elevation. The semicircle is drawn on the base AB and divided into the usual six equal parts, although only those on the quadrant from A to 3 will be needed. Again, the points on the quadrant are projected back to the base AB, and from the points on the base, lines are drawn to the apex C. Also, radial lines from points 1 and 2 are drawn to the plan apex C'.

In this case the plan of the hyperbola is a straight line, and is represented in the half plan from R to R'.

The projection of the true shape of the curve is made at right angles to OR, which in this case is horizontal, and is shown at $R''O''R''$. The points where the radial lines cross OR are projected

Polygons and Curves

horizontally, and centreline $O''X$ is drawn in any convenient position. Then, as in the previous examples, the perpendicular distances from the base AB to the points on $1,C'$ and $2,C'$ are taken and marked off on the corresponding lines on both sides of the centreline $O''X$.

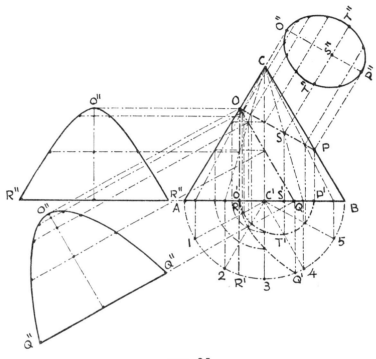

FIG. 25

Also, the full width RR', is taken from the plan and marked off on both sides of the centrepoint X. Sufficient points are thereby afforded through which to draw the hyperbolic curve, as shown in Fig. 24.

The three examples shown separately in Figs. 22, 23 and 24, might conveniently be combined into one exercise as shown in Fig. 25.

2 Projections

Projection drawing includes a number of different methods of representing objects by mechanical or instrument drawing as distinct from pictorial or freehand drawing. Among methods of projection are orthographic, auxiliary, isometric and oblique projections.

Orthographic Projection

Ordinary plans and elevations are based on orthographic projection, which generally consists of two views, a front elevation and a plan,

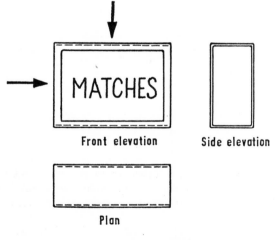

Front elevation   Side elevation

Plan

FIRST ANGLE PROJECTION

FIG. 26

though sometimes a side elevation is also included. The front elevation represents the object as seen horizontally from the front position. The plan is a view of the object as it would be seen looking

Projections

vertically downwards from above. A side elevation presents a view of the object as seen horizontally, at right angles to the front elevation, from one side or the other. Orthographic projection, therefore, presents two, three or four views in horizontal and vertical planes at right angles to each other.

Figure 26 is a simple illustration of this method of projection, in which an ordinary matchbox is shown in a front elevation, a plan and a side elevation. The relative positions of these views are

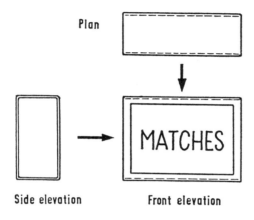

THIRD ANGLE PROJECTION

FIG. 27

important. Mistakes can easily be made if the views are not placed correctly in relation to each other.

There are in general use two methods of placing the plan and elevations in relation to each other. The British method is to place the plan vertically below the front elevation, and the side elevations on the far side of the position from which the object is viewed, as indicated by the arrows in Fig. 26.

The method more favoured in America is to place the plan and side elevations adjacent to the sides to which they refer. Thus, the plan is placed above the front elevation, and a side elevation is placed on the same side from which the object is viewed. Fig. 27 illustrates this method of projection.

The British method is generally referred to as first angle projection and the American as third angle projection. Sometimes the two methods are combined, wherein the plan is placed below the front elevation, as in first angle projection, and a side elevation is placed adjacent to the side which it is intended to represent as in third angle projection. In sheet metal work this combination can be confusing unless the direction of the views is clearly indicated. An article can easily be made left-handed instead of right-handed as intended, or vice versa. Therefore great care should be exercised in making drawings and also in reading them. In all the following examples first angle projection will be used.

Fig. 28 represents a cylindrical connecting piece to fit on a larger cylindrical surface, and is given here as an example of typical orthographic projection in sheet metal work. The construction is

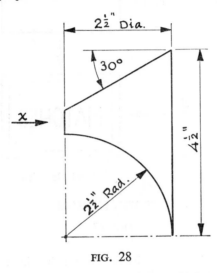

FIG. 28

shown in Fig. 29, in which the elevation is lettered $ABCD$, and a side elevation drawn as seen in the direction of the arrow X in Fig. 28. The plan of the cylinder is drawn below the elevation, and is a simple circle since the view from above is equivalent to looking down the central axis of the pipe.

To obtain the side elevation, the circle in the plan is divided into the usual twelve equal parts, and from the points on the circle vertical lines are drawn to cut the top and bottom lines AB and CD in the elevation. From the points on these two lines horizontals are drawn into the side elevation. Next, the points on the circle in the plan are drawn horizontally to meet BC produced below C. Then, using point C as centre, the points on BC produced are swung round through 90° to the horizontal base line passing through C, and from the base line the points or lines are projected vertically upwards to

Projections

cut the horizontal lines from the elevation. The appropriate points where the lines cross should give sufficient points of guidance to draw in the curves representing the top and bottom edges as shown in the diagram. The side elevation is then completed by drawing in the two outside lines between the top and bottom curves.

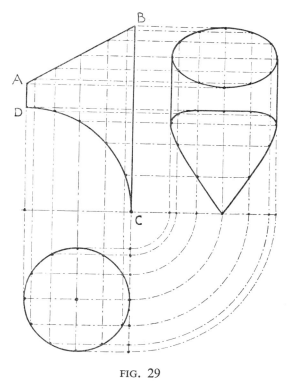

FIG. 29

Auxiliary Projection

Sometimes it is necessary to draw a view of an object as seen from an angle other than the right angles adopted in orthographic projection. This, then, becomes an auxiliary projection. An auxiliary projection is shown in Fig. 30, in which a view of the matchbox is projected in the direction of the arrow. The base or ground line $A'B'$ is made in any convenient position at right angles to the direction of projection. Then all the oblique heights from $A'B'$ are equal to the corresponding vertical heights from the base AB in the front elevation.

Auxiliary projections are often useful when it is required to obtain true angles which are not shown in orthographic plan and elevation. Some of the more difficult problems of pattern development cannot be solved without auxiliary projection.

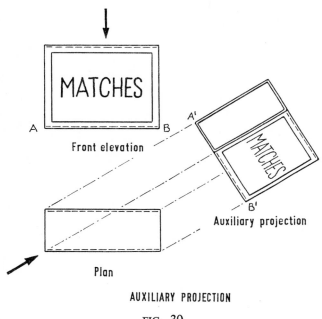

AUXILIARY PROJECTION

FIG. 30

A further example of the use of auxiliary projection is shown in Fig. 31. The orthographic plan and elevation do not show the true angle between the off-set pipe and the vertical portions. In order to obtain the true off-set view, an auxiliary projection is made at right angles to the plan centreline, as at $A'B'C'$. In the auxiliary projection the height $B'C'$ is made equal to the height BC in the elevation. The length $A'C'$ in the projection then presents the true length of that portion of the centreline, and the angles $D'A'C'$ and $A'C'E'$ are the true angles between $D'A'$ and $A'C'$ and between $A'C'$ and $C'E'$. The joints, then, at A' and C' will present straight lines, and the patterns may readily be developed from that view.

Projections

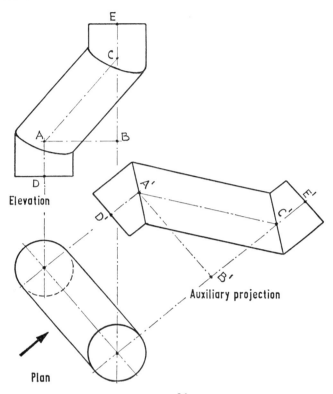

FIG. 31

Isometric Projection

There are numerous methods of drawing objects to represent them pictorially. One important method is that of isometric projection. In this method all lines which would normally occur horizontally in the front and side elevations are drawn at 30° to a horizontal ground line, as shown in Fig. 32. All vertical lines are drawn vertically. It is usual to make the lengths of the lines conform to a scale of the full dimensions. This method is not one of true perspective, nor does it present true dimensional accuracy, but is a good approximation in which the lines may be drawn with ordinary set squares.

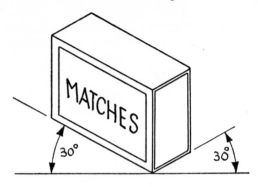

ISOMETRIC PROJECTION

FIG. 32

The illustration in Fig. 33 shows a hood, or transforming piece, in orthographic projection. The same hood is shown in Fig. 34 in isometric projection.

The centre of the circular hole at the top is vertically above the centre of the rectangle at the bottom. One convenient method of

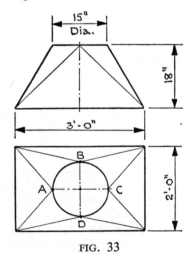

FIG. 33

obtaining this condition in the isometric projection is to repeat the rectangle at the given height above the bottom rectangle, as shown in chain dotted lines in Fig. 34, and then locate the position of the

Projections

circle, which in the isometric view will become an ellipse, in the centre of the top rectangle. To obtain the ellipse, the diameter of the circle is marked off on each of the crosslines, as at *AC* and *BD*, and the

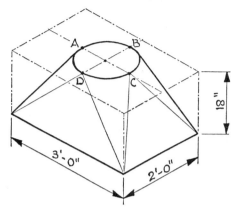

FIG. 34. *Isometric view of transforming piece*

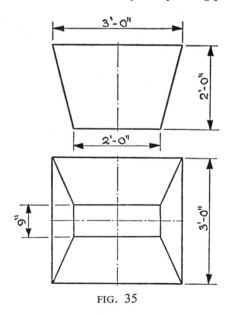

FIG. 35

ellipse drawn through these points. It will be seen, however, that this results in the ellipse presenting a larger diameter than that of the circle, which in a true projection would be impossible.

The hopper shown in orthographic projection in Fig. 35 is shown in isometric projection in Fig. 36. A similar method is used in this case for locating the position of the bottom in relation to the top, inasmuch as the shape of the square top is repeated at the given depth below the top. The rectangular bottom is then located in its correct position within the square. The lines representing the corners may then be drawn in as shown in the Figure.

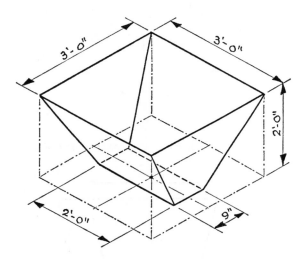

FIG. 36. *Isometric view of hopper*

Oblique Projection

The chief difference between oblique projection and isometric projection is that in oblique projection one view is drawn as a normal elevation, as, for example, the end elevation of the box shown in Fig. 37, and then the depth lines are added at an angle of 30° or 45° on one side of the elevation, as shown in Figs. 38 and 39.

The front view of an oblique projection may be either a normal front elevation or a side elevation, or even a cross-section if desired. It is not essential that an angle of 30° or 45° should be used for the depth lines, as almost any other angle would give satisfactory results, but these angles lend themselves readily to the use of the common set squares.

Projections

One drawback to oblique projection is that if the depth lines were made to the given dimensions, the object would appear to be larger than it really is.

These methods of projection, isometric and oblique, present the object pictorially, though generally they are not accurate representations. On the other hand, auxiliary projections, if correctly drawn, should present accurate views of the object as seen in the direction chosen at any desired angle.

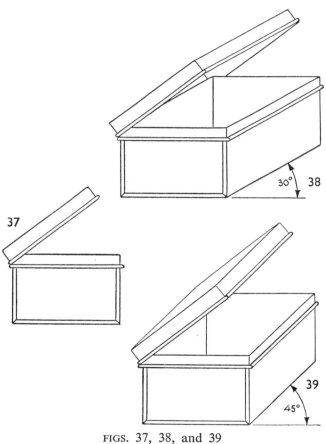

FIGS. 37, 38, and 39

3 Radial Line Development

The pattern for any article which tapers to an apex may be developed by the radial line method. This method also applies to frustums which would taper to an apex if the sides were produced.

The radial line method is based on the location of a series of lines which radiate from the apex down the surface of the object to a base, or assumed base, from which a curve may be drawn on which the perimeter of the base may be marked off.

A RIGHT CONICAL HOOD

The example shown in Fig. 40 represents a right conical hood with a pipe elbow which conveys the fumes through a wall to outside atmosphere. The development of the pattern for the hood is shown

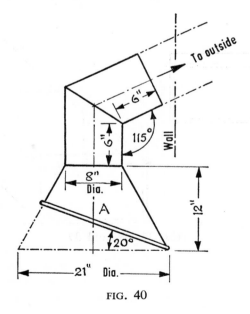

FIG. 40

Radial Line Development

in Fig. 41. The sides of the hood are produced upwards to the apex *A*, and the assumed base of the cone *BC* is drawn at right angles to the centreline *AG*. The base *BC* is therefore circular.

The semicircle drawn on the base *BC*, representing half its plan, is divided into six equal parts and the points projected perpendicularly back to *BC*. From the points on the base, lines are drawn

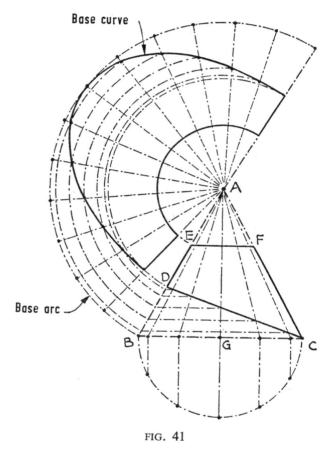

FIG. 41

to the apex *A*, and from the points where these lines cross the inclined base *DC*, lines are projected horizontally to the side of the cone. With apex *A* as centre, the base arc is drawn from point *B*. Then one of the equal spacings is taken from the semicircle on *BC*, and twelve of these spacings are marked off along the base arc, commencing at any convenient point. Lines are now drawn from the points thus obtained on the base arc to the apex *A*.

Next, with apex *A* as centre, a series of arcs are drawn from the points on the side of the cone between *B* and *D*, and also an arc from point *E*. Where the series of arcs cross the radial lines, points are afforded through which to draw the base curve in the pattern. The full outline may then be completed as shown in Fig. 41.

A RIGHT CONICAL CONNEXION TO ANGULAR SURFACE

The right conical connexion shown in Fig. 42 presents a further example of simple radial line development. To develop the pattern,

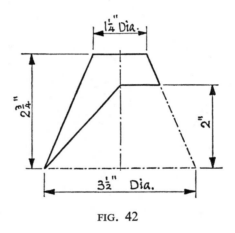

FIG. 42

the elevation should be drawn as shown in Fig. 43, and the sides *AG* and *BF* produced to the apex *O*. Then a semicircle is described on the base *AB* of the cone, and the left-hand quadrant *AE* divided into three equal parts. Next, the points on the quadrants *AE* are projected vertically upwards to the base of the cone, and from the points on the base, lines are drawn to the apex *O*. From the points where the lines to the apex cross the cut-off *AC*, lines are drawn horizontally to the side of the cone. These points on the side of the cone represent true lengths from the apex, and each one is now swung into the pattern, using the apex *O* as centre. Next, the outermost curve from *A* is divided into twelve parts equal to those on the quadrant *AE*. The first and last points and the middle six divisions are now joined to the apex *O*. Where these radial lines cross the curves from the side of the cone, sufficient points are afforded to

Radial Line Development

draw in the pattern as shown in the diagram. The seam in this case is arranged to be at *FD* on the short side of the cone.

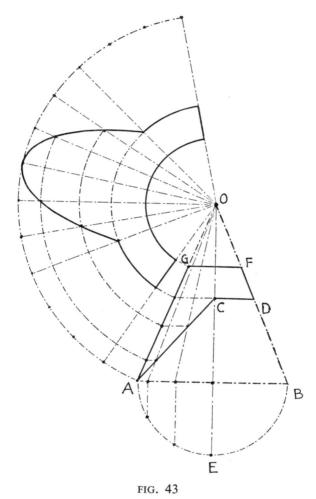

FIG. 43

A RIGHT CONICAL OUTLET

The example shown in Fig. 44 represents a right conical outlet from a cylindrical duct. To develop the pattern for the outlet, a semicircle is first drawn on the base *BC*, as shown in Fig. 45, and

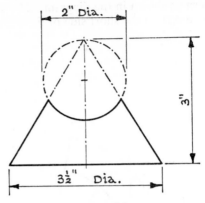

FIG. 44

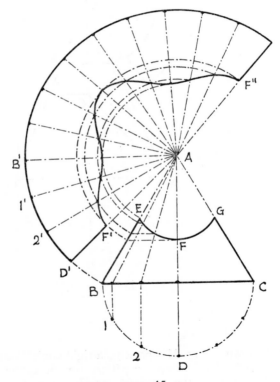

FIG. 45

Radial Line Development

35

divided into the usual six equal parts. Only three of these parts, between *B* and *D*, will be needed in the development. The points 1 and 2 are projected vertically upwards to the base *BC*, and from the points on the base, lines are drawn to the apex *A*. From the points where these and the centreline cut the arc *EFG*, lines are projected horizontally to the outside *AB* of the cone. From all the

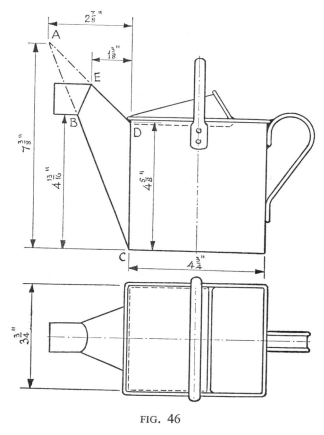

FIG. 46

points on *AB*, including the base point *B*, arcs are swung into the pattern.

Next, one of the equal parts is taken from the quadrant *BD*, and twelve equal parts marked off along the outside arc as shown in Fig. 45. From the points on the outside arc, lines are drawn to the apex *A*. Where these lines cross the remaining arcs, points are afforded through which to draw the joint curve $F'F''$ in the pattern. It will be noted that the position of the joint is arranged to occur on the short side as at F.

A SPOUT ON A RECTANGULAR CAN

Fig. 46 presents a right conical spout, $BCDE$, on the flat surface of a rectangular body. The pattern for the conical part of the spout is shown developed in Fig. 47.

Since the spout $BCDE$ is a portion of a right cone, the sides BC and DE produced meet at the apex A. The side AD is next produced

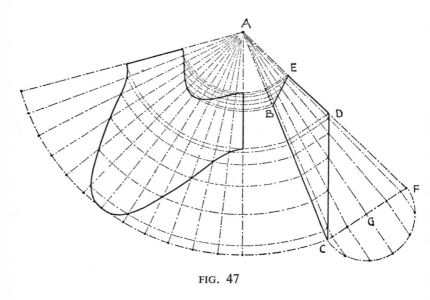

FIG. 47

to a point F such that AF is equal to AC. Then CF serves as the base of the cone, and is bisected in G to obtain the centreline AG of the cone.

Next, a semicircle representing half of the base of the cone is drawn on CF, and divided into six equal parts. The points on the semicircle are now projected perpendicularly back to the base line CF. Then from the points on CF, lines are drawn to the apex A. From the points where these lines cut CD and BE, lines are projected parallel to the base CF, or at right angles to the centreline AG, to meet the side AC of the cone.

Now, using A as centre, arcs are swung into the pattern from all the

Radial Line Development

points on the side AC. Next, one of the equal spacings is taken from the semicircle on CF, and twelve of these spacings are marked off along the outside arc. Lines are then drawn from the points on the arc to the apex A. Assuming that the joint in the conical body is to be on the upper side DE, the two curves in the pattern are drawn as shown in Fig. 47.

A WATERING-CAN ROSE

The example shown in Fig. 48 represents a rose for a watering-can. The body of the rose is a frustum of a right cone, and the top is an elliptical piece of metal slightly hollowed and perforated.

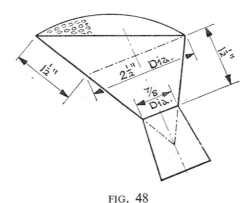

FIG. 48

For the development of the cone, the body should be set out to the given dimensions as shown in Fig. 49. As the diameter is given at the $2\frac{1}{2}$-in. position at right angles to the central axis of the cone, a semicircle may be drawn on that diameter, as at BC, and divided into the usual six equal parts. The points on the semicircle are then projected back to the diameter, and lines drawn from the apex A through the points on the diameter to the edge CD of the cone. Next, from the points on CD, lines are pojected parallel to BC to the side AD of the cone. Then with centre A, arcs are swung into the pattern from all the points on AD. As the semicircle is drawn on the diameter BC, one of the divisions on the semicircle is taken and spaced off along the arc drawn from point B. Radial lines from the apex A are now drawn through the points on the arc to cut the remaining arcs. Points are thus obtained on the arcs through which to draw the curve in the pattern as shown in Fig. 49.

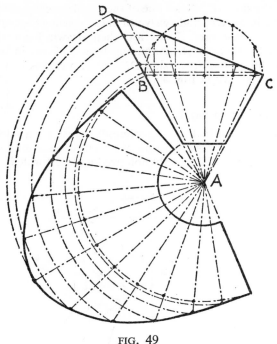

FIG. 49

AN EQUAL-TAPERED TRAY

Shallow tapered trays are a fairly common utilitarian type of article usually made of tinplate. Most of them are of simple square or rectangular construction, though other shapes are sometimes required for special purposes. One such tray is illustrated in Fig. 50, which shows a tray of equal taper or constant angle of slope all round the body.

From a geometrical point of view this tray is a composite article, and the pattern development needs to be sectionalized according to the construction. Taking the half plan from A to A^0, Fig. 51, a short straight side Ab, is connected to a quarter of a right cone bc, followed by a further straight side cd, which in turn is connected to a quarter of an oval or elliptical equal-tapered portion dA^0. It is the elliptical portion which needs a little careful consideration before deciding on the method of development. Since it is to be equal tapered this portion cannot have a common apex, or the side and end would meet the central axis at different heights. This problem

Radial Line Development

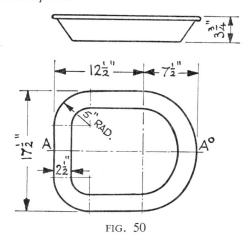

FIG. 50

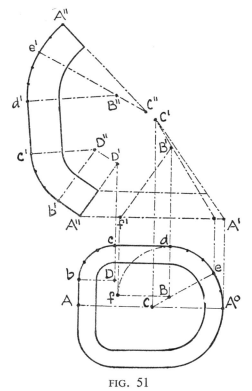

FIG. 51

is illustrated in Fig. 52, where the plan and elevation of an elliptical pyramid are shown. It will be seen that if the apex were at the same height in the front and the side elevations, then the angle of taper would be different in the two views. Alternatively, if the taper is to be equal all round, then the apparent apex in the side view is at a lower point than that in the front view. Therefore in the latter case the object has no common apex and the pattern cannot be developed as a simple radial line problem. It could, however, be developed quite easily by the method of triangulation described in Chapter 6.

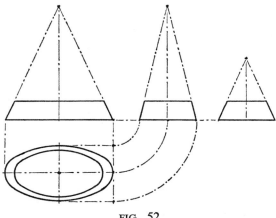

FIG. 52

Assuming that the plan view, instead of being a true ellipse, were drawn as an oval by connecting the arcs of circles as shown in Fig. 53, then the pattern could be developed by combining two portions of right cones having different heights and base diameters. In Fig. 53, a right conic portion forming the ends of the body is shown with the apex at B in the plan and at B' in the elevation. Also, the apex of a right conic portion forming the sides of the body is shown at C in the plan and at C' in the elevation.

The development of the pattern for the tray is shown in Fig. 51, based on the right conic portions indicated above. To develop the pattern, first set off the straight portion $A''b'$ equal in length to Ab, and in width equal to the slant of the pan. Extend A'' to D'. The point D' represents the apex of the quarter cone bDc, and $A''D'$ the side of the cone. Transfer D' to D'', making $b'D''$ equal to $A''D'$, then D'' becomes the centre for the quarter conic pattern $b'D''c'$. Add the next straight portion $c'd'$, making it equal in length to cd in the plan. Extend d' to B'', making $d'B''$ equal in length to $f'B'$, which

Radial Line Development

is the slant side of the conic portion dBe. From the centre B'', draw the arc $d'e'$, making it equal in length to the arc de in the plan. Extend $e'B''$ to C'', making $e'C''$ equal in length to $C'A'$, which is the slant side of the conic portion eCA^0. From the centre C'' draw the arc $e'A''$, making it equal in length to the arc eA^0 in the plan. This completes the outside curve of the pattern. The inside curve should

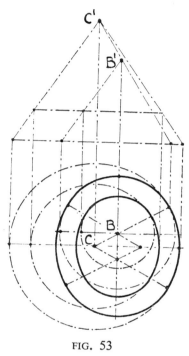

FIG. 53

maintain a constant distance from the outside curve equal to the slant side of the pan.

THE OBLIQUE CONE

The methods of developing patterns for oblique cones and their frustums appear to be quite different from those applied to the development of right cones and their frustums. Nevertheless, the basic principles are the same.

Referring to Fig. 54(a), which represents an oblique cone with the apex at A, the line AB represents a line drawn to the apex A from any point B on the base. Point C represents any point on the line

AB. For purposes of pattern development it is essential to determine the true distance of *C* from the apex *A*. The elevation line *AB* in its given position does not present its true length, as the apex leans backwards from the point *B*. In order to present the true length of *AB*, its plan length *A'B'* must be placed at right angles

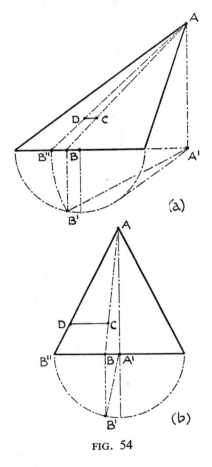

FIG. 54

to its vertical height *A'A*, when the diagonal *AB"* will give the true length. Thus, from the point *A'*, which represents the plan of the apex *A*, the point *B'* is swung round to point *B"* on the base of the cone. Then the line *AB"* represents the true length of the elevation line *AB*. It may be readily seen that the true distance from *A* of any

Radial Line Development

point C on AB can be obtained by transferring C horizontally to D on AB''. Then AD represents the true length of AC.

Now compare this with a similar condition on a right cone, Fig. 54(b). The plan of the apex A in this case falls at the centre A' of the semicircle on the base. The point B on the base presents a plan line from A' to B'. If, now, the plan line $A'B'$ is swung round at right angles to the vertical height AA', then the point B'' is obtained on the outside point of the base of the cone. Thus, in this case, the outside line AB'' in the elevation of the cone, represents the true length of the elevation line AB. Then the true distance from A of any point C on the elevation line AB may be obtained by projecting point C horizontally to the outside line in the elevation of the cone.

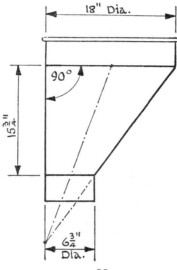

FIG. 55

By the same line of reasoning it will be seen that the true distance from the apex to any point on the surface of a right cone may be found by projecting the point horizontally to the outside line of the cone. Comparing this with similar conditions on the oblique cone, it will be found that the projection of any point C on the surface of the cone must be made horizontally to a corresponding true length line. In the case of the oblique cone, the true length lines will not occur on the outside line of the cone, but will differ according to the position of the elevation line passing through C. The fundamental principle, therefore, is the same in both cases, though the methods of development appear to be different.

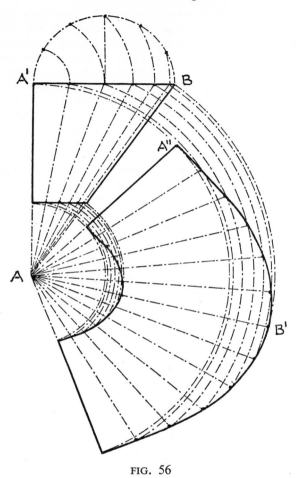

FIG. 56

AN OBLIQUE CONICAL HOPPER

In Fig. 55 an elevation is given of a hopper to be made of sheet metal. The construction of the hopper is a frustum of an oblique cone.

The development of the pattern is shown in Fig. 56. In this case the base of the cone is at the top and the apex below. As the angle of the cone at A' is 90°, the apex A is vertically below A'. The usual semicircle is described on $A'B$ to represent half of the base of the cone, and divided into six equal parts. Then, with centre A', the

Radial Line Development

points on the semicircle are swung round to the base $A'B$. The points on $A'B$ are then swung into the pattern with the apex A as centre. Next, one of the equal spacings is taken from the semicircle, and beginning at any point A'' on the innermost of the arcs drawn from $A'B$, the spacings are stepped over from one arc to the next, until the outer arc is reached at B'. The spacings are then continued back to the inner arc. This method of spacing from one arc to the next ensures that the length of the base curve in the pattern is the same as that of the circumference of the circular base of the cone at $A'B$.

To obtain the inner curve of the pattern, lines are drawn from the points on the base $A'B$ to the apex A. Where these lines cross the smaller edge of the cone, points are afforded from which to swing arcs into the pattern from the apex A. Lines are now drawn to the apex A from the points on the base curve $A''B'$. Where these lines cross the arcs drawn from the smaller edge of the cone, points are afforded through which to draw the inner curve of the pattern, as shown in Fig. 56.

A NON-SYMMETRICAL NOZZLE

The fishtail nozzle shown in Fig. 57 is an example of a composite body combining two halves of different oblique cones. One half cone may be seen in the plan, Fig. 58 with the apex at A and the half-circular base from point 1 to point 7. The corresponding half-circular top occurs at CD. The other half cone has its apex

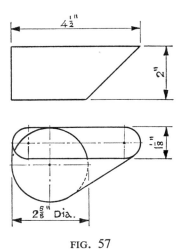

FIG. 57

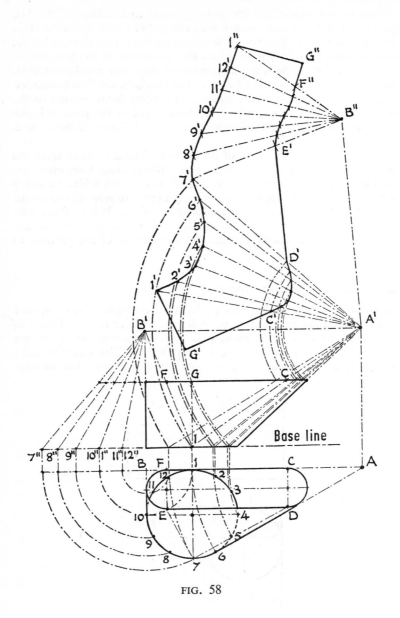

FIG. 58

Radial Line Development

at B, and the half-circular base from point 7, rotating clockwise, back to point 1. The corresponding half-circular top lies between points E and F. Two flat triangles, $D7E$ and $F1C$, join the two half cones together, forming the complete fishtail nozzle.

The full pattern is developed as shown in Fig. 58, with the seam vertically downwards at the position of point 1. The first part of the pattern to be considered is that of the right-hand semi-cone, in which the two sides of the frustum in the plan are produced to locate the apex at A. This apex is projected vertically upwards to meet the slant of the frustum in A', which represents the apex in the elevation. Next, the semicircle from points 1 to 7 is divided into six equal parts and numbered as shown in Fig. 58. Then, using A as centre, the points on the semicircle are swung round to the horizontal line AB, and from AB the points are continued vertically upwards to the base line in the elevation. Now, using the apex A' as centre, the points thus obtained on the base line are swung out into the pattern. Next, one of the equal divisions is taken from the semicircle in the plan, and, beginning on the arc obtained from point 1, this distance is stepped over from one arc to another. The next arc in consecutive order must correspond to that obtained from the next point around the semicircle; that is, from 1 to 2, 2 to 3, 3 to 4, and so on. This should be readily followed by inspecting the points from 1' to 7' in the pattern shown in Fig. 58.

The next part of the development is to obtain the inside curve in the pattern corresponding to the top edge of the cone. The first step is to draw lines from the points on the base to the apex A'. Then, using A' as centre, swing arcs into the pattern from the points where those lines cross the top edge in the elevation. Next, from the points 1', 2', 3', 4', 5', 6' and 7', lines are drawn to the apex A'. Where these lines cross the arcs from the top edge, points are afforded through which the inside curves may be drawn. It is important that the points on the inner curve should occur on the arcs which correspond to those on the outer curve.

The other half oblique cone with the apex B in the plan and B' in the elevation is the next part to be dealt with in the development. With the apex B as centre, each of the points 7, 8, 9, 10, 11, 12 and 1 is swung round to the extended line AB, and the points thus obtained on that line are then projected vertically upwards to the extended base line in the elevation, as shown in Fig. 58. From the points on the base line, lines are drawn to the apex B'. These lines are the true lengths of the plan distances from the apex B to the points around the semicircle. The plan lines are not shown.

Now, to proceed with the pattern, the position of the apex B'' is first located by making the distance $A'B''$ equal to $A'B'$, and the distance $7'B''$ equal to the true length $7''B'$. Then, the true distances $B'8''$, $B'9''$, $B'10''$, $B'11''$, $B'12''$ and $N'1''$ are taken in turn from the elevation and marked off in arcs from the apex B'' in the pattern. The spacings $7'8'$ to $12'1''$ are made equal to the divisions around the semicircle in the plan from 7 to 1, and lines are drawn from those points to the apex B''. The inner curve may now be plotted by taking the true length distances from the apex B' in the elevation to the top edge produced, and marking these off from the apex B'' in the pattern. It is important that these distances should be marked on the corresponding lines radiating from B'' to the outer curve. Next, the points D' and E' are joined to form the flat triangle $D'7'E'$.

Finally, the two triangles, $1'C'G'$ and $1''F''G''$, are added to complete the pattern. These triangles may be taken direct from the elevation, in the one case making $1'G'$ equal to $1,G$ and $C'G'$ equal to CG, and in the other case making $1''G''$ equal to $1,G$ and $F''G''$ equal to FG.

AN OBLIQUE CONICAL HOOD

The illustration in Fig. 59 is of an oblique conical hood which could serve the purpose, in a system of ductwork, of drawing off steam from a boiling kettle. In leaning backward, this type of hood leaves a large part of the top of the kettle free for filling and stirring operations; at the same time an adequate updraught draws the steam across the top of the kettle.

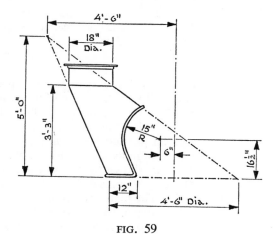

FIG. 59

Radial Line Development

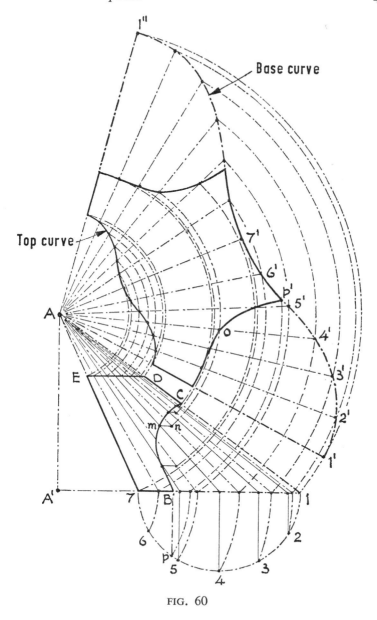

FIG. 60

The development of the pattern for the hood is shown in Fig. 60. The pattern for the full conic frustum is first obtained, and then the shape of the curved front edge located on the radial lines in the pattern. The semicircle is drawn on the base of the cone and divided into the usual six equal parts, as from 1 to 7. Next, the apex A is dropped perpendicularly to A' on the base 1,7 produced. The point A' then represents the position of the apex in the plan. Now, with point A' as centre, the points 2, 3, 4, 5 and 6 on the semicircle are swung round to the base line 1,7. Then, from the points thus obtained on the base, lines are drawn to the apex A. These lines are *true length lines*, and from the apex A, arcs are swung into the pattern. Next, one of the spacings is taken from the semicircle and, beginning at a convenient spot on the outside arc as at $1'$, the spacings are stepped over from one line to the next until the inside arc is reached at $7'$. The spacings are then continued, stepping outwards, until the outside arc is again reached at point $1''$. The base curve is then drawn in through the points on the arcs, as shown in the illustration.

From the points on the base curve, radial lines are drawn to the apex A. Next, another series of arcs are drawn from the top edge ED of the hood. Where the arcs cross the radial lines, points are afforded through which to draw the top curve in the pattern. Note that the top curve follows a similar form to that of the base curve.

It now remains to plot the shape of the curved edge, BC, in the pattern. First, from the points 2, 3, 4 and 5 on the semicircle, lines are drawn perpendicular to the base 1,7. From the points obtained on the base, lines are drawn to the apex A. Note that these lines are *elevation lines*, as distinct from the true length lines previously drawn, also that an elevation line and a true length line are obtained from each of the points 2, 3, 4 and 5. Now, where each elevation line crosses the curved edge BC, the point is projected horizontally to the corresponding true length line. Then, the distances from the apex A to the points on the true length lines are swung into the pattern to meet the corresponding radial lines from the base curve.

For example, the elevation line from point 4 on the semicircle cuts the curve BC in point m. Point m is then projected horizontally to the true length line which also originates from point 4 on the semicircle, to obtain point n. The distance An is then swung into the pattern to obtain point o on the radial line from point $4'$. This process repeated with the other elevation and true length lines will provide a number of points on the radial lines in the pattern. The last point, however, still needs to be located on the base curve, as at p'. From the point B on the base 1,7, a perpendicular line is dropped to the semicircle to obtain the point p. The short distance

Radial Line Development

from point 5 to point p is then taken and marked off along the pattern curve from point $5'$ to point p'. Point p' now completes the series through which the curved edge of the hood may be drawn in the pattern. A similar curve on the opposite side of the centreline completes the shape of the pattern for the hood.

The extra point p' in the pattern may alternatively be obtained by treating the plan point p in a similar manner to the other points on the semicircle. That is, from the centre A', the point p is swung round to the base of the cone, and then from the apex A, the point on the base is swung into the pattern to give the extra curve as shown in the Figure. Next, from the plan semicircle the true distance $6,p$ is taken and from point $6'$ in the pattern an arc is drawn cutting the extra curve in point p'. This method may be used in all similar conic problems where an extra point needs to be obtained on the base curve in the pattern.

AN OBLIQUE CONICAL HOPPER

The hopper shown in Fig. 61 is made to fit on a flat-topped body with a corner radius. The hopper is of oblique conical construction with a wired top and a flanged bottom for fixing to the body.

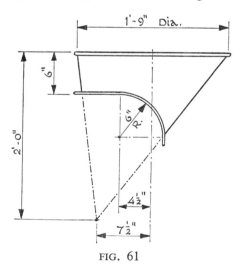

FIG. 61

The pattern for the hopper is shown developed in Fig. 62. In this example the base is inverted, and the apex A falls upwards to the point A' on the base line 1,7. The semicircle representing half the inverted plan is drawn on the base line, divided into six equal parts, and numbered as from 1 to 7. To obtain true length lines for the pattern, using A' as centre, each of these points is swung round to the base line 1,7. Then, using the apex A as centre, each of the points obtained on the base line is swung into the pattern. Next, one of the equal spacings is taken from the semicircle, and, beginning on the inside curve, the distance is spaced over from one arc to the next, as shown in the Figure, to obtain the base curve from $1'$ to $7'$. From the point $7'$ on the outside curve, the spacing is continued in the reverse order back to the inside curve. From each of the points on the base curve, lines are now drawn to the apex A.

The next stage is to obtain the joint curve in the pattern. To do this, elevation lines and true length lines must first be drawn in the elevation. The elevation lines are obtained by dropping perpendiculars from the points on the semicircle to the base line 1,7, and then from the points on the base, lines are drawn to the apex A. These are *elevation* lines, shown in full line. The true length lines are next obtained by drawing lines to the apex A from the other set of points on the base line 1,7. These points are the ones where the arcs from the semicircle fall on the base line, and have already been used when swinging the larger arcs into the pattern. The lines now drawn from these points to the apex A are *true length* lines, shown chain dotted. Thus, from each of the points 2, 3, 4, 5 and 6 on the semicircle, an elevation line and a corresponding true length line are obtained in the elevation.

Now, where each of the elevation lines crosses the curved edge BC, a horizontal line is drawn to the corresponding true length line. The points thus obtained on the true length lines are then swung into the pattern to meet or cross the corresponding radial line from the apex A. For example, the elevation line obtained from point 3 on the semicircle crosses the curved edge in point m. A horizontal line drawn from point m meets the corresponding true length line in point n. Then, with A as centre, an arc is swung from point n into the pattern to cross the radial line running from A to point $3'$. The arc is continued to meet the corresponding radial line on the other side of the pattern. As a further example, the elevation line from point 4 on the semicircle crosses the curved edge in point o, and the horizontal line from point o meets the corresponding true length line in point p. Then, from point p an arc is swung into the pattern to cross the radial line running from A to point $4'$, and continued to meet the corresponding radial line on the other side of the pattern. This process, repeated with the other pairs of elevation and true length lines, provides sufficient points in the pattern through which to draw the joint curve, as shown in Fig. 62.

Radial Line Development

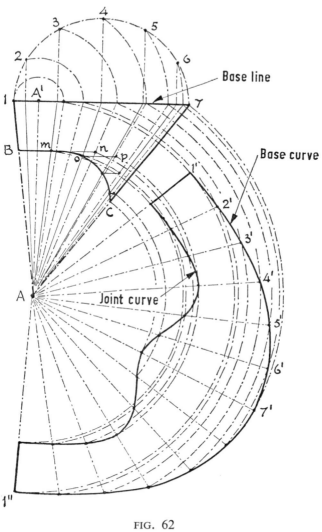

FIG. 62

A THREE-WAY CONICAL BRANCH PIECE

The three-way branch piece shown in Fig. 63 is composed of three conic frustums; the middle one is a right cone and the two outside ones are of oblique conic construction. The development of the patterns is shown in Fig. 64, in which the middle right cone and the right-hand oblique cone are set out for development.

The semicircle is described on the common base, divided into six equal parts, and numbered as from 1 to 7. To deal with the right cone pattern first, lines are drawn from points 2 and 3 vertically

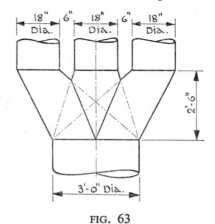

FIG. 63

upwards to the base, and from these points on the base, lines are drawn to the apex A. Then, from the points where these lines cross the left-hand joint line, lines are drawn parallel to the base as far as the side of the cone. Using the apex A as centre, arcs are now swung into the pattern from all the points on the side of the cone. The base arc is now divided into twelve parts equal to those around the semicircle, half of which are numbered from $1'$ to $7'$ in the Figure. From all the points on the base arc, radial lines are drawn to the apex A. Where the radial lines cross the arcs from the side of the cone, points are afforded through which to draw the joint curve in the pattern as shown in Fig. 64.

Now, to deal with the pattern for the oblique cone, the apex A' is dropped perpendicularly to the extended base 1,7 to obtain the plan apex A''. Using the apex A'' as centre, the points 1, 2, 3, 4, 5, 6 and 7 on the semicircle are swung round to the base line 1,7. From the points thus obtained on the base, lines are drawn to the apex A'.

Radial Line Development

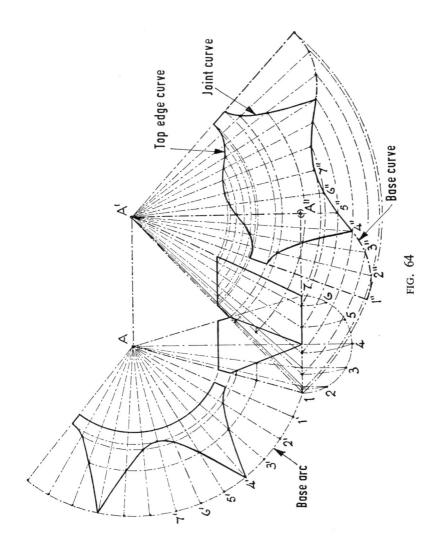

FIG. 64

These are *true length* lines. Then, using the apex A' as centre, the true length lines are swung round from the base into the pattern. Also, where the true length lines cross the top edge of the cone, those points are swung into the pattern.

Next, one of the equal spacings is taken from the semicircle, and beginning at a convenient point on the outside arc, the spacing is stepped over from one arc to the next until the inside arc is reached. The spacing is then reversed back to the outside arc. The first half of these spacings is numbered as from $1''$ to $7''$ in the illustration. A line drawn through these points gives the shape of the base curve in the pattern. Now, from each of the points on the base curve, radial lines are drawn to the apex A', and where the radial lines cross the arcs from the top edge of the cone, points are afforded through which to draw the top edge curve in the pattern.

It now remains to determine the joint curves in the pattern. It should be observed that the bottom point of the joint line occurs at point $4''$, and the top point of the joint line occurs on the first radial line on the arc described from the top of the joint line in the elevation. The true distances from the apex A' of the other two points on the radial lines $2''$ and $3''$ have first to be obtained in the elevation. Thus, from the points where the respective elevation lines cross the joint line, short horizontal lines are drawn to the corresponding true length lines. Then from the apex A', the points on the true length lines are swung into the pattern to cut the radial lines $2''$ and $3''$ and also the corresponding radial lines on the other side of the pattern. The joint curves are now drawn through those points as shown in Fig. 64.

AN OBLIQUE CONICAL FOUR-WAY BRANCH PIECE

The four-way branch piece shown in Fig. 65 is a further example of oblique conic construction. Since the tops of the branches are circular and parallel with the circular base, each branch forms a frustum of an oblique cone. The development of the pattern is shown in Fig. 66. Since these four branches are oblique conic frustums, each one would, if completed as a single frustum, taper from the circular base of 3 ft diameter to the circular top of 1 ft 9 in. Therefore, in setting out the problem for development, the elevation is drawn as shown in Fig. 66, and the joint line obtained in accordance with the position of the cut-off shown in the plan. To determine the joint line in the elevation, first complete the cone by producing the sides of the cone to the apex A'. Drop the apex into the plan to obtain the apex A. Next, divide the circular base BC into twelve equal parts and join the points to the apex A. Also

Radial Line Development

project the points on the circular base vertically upwards to the base line $B'C'$ in the elevation, and join these points to the apex A'. Thus, the radial lines in the plan are also represented in the elevation. Now, in the plan, these lines are cut by the joint lines OE and OD, and where the lines cross the joint line OE, the points are projected vertically upwards to the corresponding lines in the elevation. The

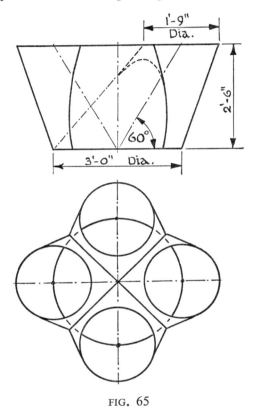

FIG. 65

points obtained on the elevation lines then provide sufficient points through which to draw the joint line as shown in Fig. 66.

Prior to developing the pattern, true length lines must be obtained in the elevation. To do this, all the points on the semicircle BDC in the plan are swung round to the centreline AB by using the apex A as centre. Then, from the centreline AB the points are projected vertically upwards to the base line $B'C'$ in the elevation. Next, from the points obtained on $B'C'$, lines are drawn to the apex A'.

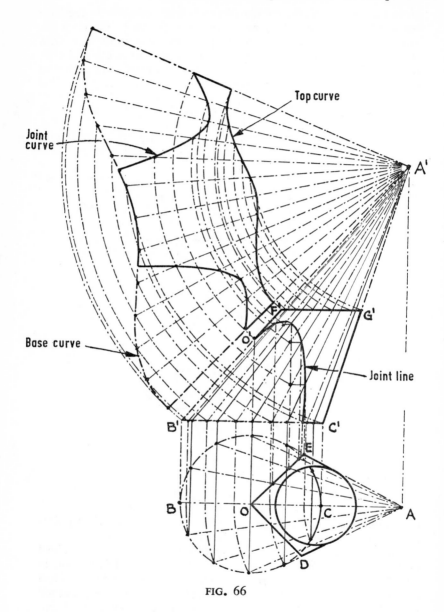

FIG. 66

Radial Line Development

These lines are now the true lengths of the corresponding elevation lines, and are represented by chain dotted lines to distinguish them from the full elevation lines.

To develop the pattern, the true length lines are swung into the pattern using the apex A' as centre. Also using the apex A' as centre, the portions of the true length lines are swung into the pattern from the points where they cut the top edge $F'G'$. Next, one of the equal divisions from the base circle in the plan is taken, and, beginning on the outside arc in the pattern, the spacings are stepped over from one arc to the next until the inside arc is reached. The spacing is then continued in the reverse order back to the outside arc. The base curve in the pattern is then drawn in, and radial lines from the points on the arcs are drawn to the apex A'. Where the radial lines cross the inside arcs from the top edge $F'G'$, points are afforded through which to draw the top curve in the pattern.

The top curve and the base curve represent the shape of the pattern for the full frustum $B'C'G'F'$. To determine the shape of the joint lines in the pattern, true lengths from the apex A' in the elevation must be swung round to the corresponding true length lines in the pattern. To obtain the true lengths in the elevation, the points where the elevation lines cross the joint line are projected horizontally to meet the corresponding true length lines. The points obtained on the true length lines give the required true distances from the apex A', and are swung round to the corresponding true length lines in the pattern. Thus, the points are now obtained through which to draw the two joint curves as shown in Fig. 66.

A CONICAL HOPPER ON CONVEYOR CASING

The illustration in Fig. 67 shows a right conical hopper fitting on a rectangular conveyor casing. The hopper may, of course, be either welded direct to the casing or suitably flanged and fixed by rivets or bolts. As the vertical sides of the conveyor are parallel to the central axis of the cone, the shape of the joint on the sides is that of a hyperbolic curve.

The development of the pattern is shown in Fig. 68. First, the left-hand elevation is drawn to the dimensions given, and the pattern for the body of the cone may be developed from that view alone. The semicircle is drawn on the inverted base of the cone, and divided into the usual six equal parts. In this example it is an advantage to further subdivide the middle divisions on each quadrant, as between 2 and 4 on the one side and between 6 and 8 on the other. These extra divisions provide an additional point on each of the vertical sides, as at c and g.

With the extra divisions on the semicircle, the points are numbered as shown from 1 to 9. Then, from the points, lines are dropped perpendicularly to the base 1,9, and from the points on the base, lines are drawn to the apex *A*. Where the lines from the points 2, 3 and 4 cross the vertical side of the conveyor at *b*, *c* and *d*, lines are drawn horizontally to the side of the cone. Using the apex *A* as centre, all the points on the side of the cone, including the base point 1, are swung into the pattern. The base arc is now divided into a number of parts similar to those on the semicircle, but in this case,

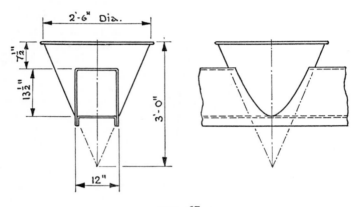

FIG. 67

instead of beginning at point 1, the divisions are started at point 5 in order that the joint shall occur on the short side. Thus, the base arc in the pattern is numbered 5'4'3'2'1', and so on as shown in the Figure. From the points on the base arc, radial lines are drawn to the apex *A*, and where the radial lines cross from the side of the cone, points are afforded through which to draw the joint curve as shown in Fig. 68.

The contour of the joint in the conveyor casing is also shown in Fig. 68. The first step in this development is to draw the side elevation shown on the right-hand side of the Figure. The two vertical sections of the cone which fit against the sides of the conveyor casing at *ad* and *fi* present hyperbolic curves in the side elevation.

When the elevation of the cone is drawn, a semicircle is described on the base, and divided into a number of parts similar to those in the left-hand elevation. From the points on the semicircle, lines are dropped perpendicularly to the base of the cone and from the points thus obtained on the base, lines are drawn to the apex *A'*. Next, lines are projected horizontally from points *f*, *g*, *h* and *i* on

Radial Line Development

the vertical section in the left-hand elevation to meet or cross the corresponding conic lines in the right-hand elevation. Points are thereby afforded through which to draw the hyperbolic curve as shown in Fig. 68. The contour of the joint line in the conveyor casing may now be developed.

Consider first the rectangular form, *adfi*, in the left-hand elevation. This represents the length of the centreline of the hole or the joint

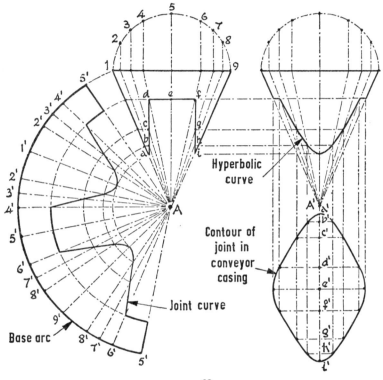

FIG. 68

area. If this were opened out flat, the portion from *a* to *d* would be a hyperbolic curve similar to that in the right-hand elevation; the portion from *d* to *f* would be similar to a cross-section of the cone at that position; and the portion from *f* to *i* would again be a hyperbolic curve similar to that in the right-hand elevation.

To obtain the full contour, the centreline of the cone in the right-hand elevation is extended downwards, and the spacings *a–b–c–d–e–f–g–h–i* taken from the left-hand elevation are marked off down

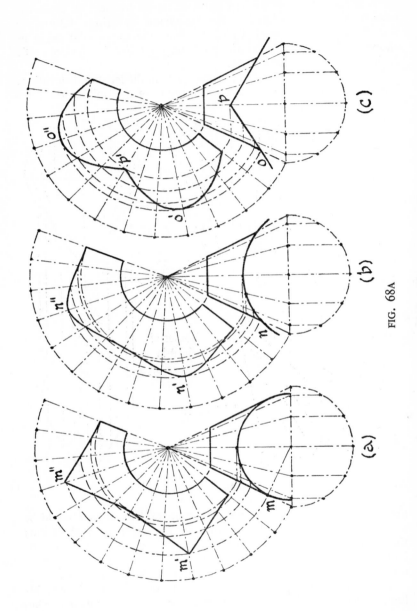

FIG. 68A

Radial Line Development

the line as shown at $a'-b'-c'-d'-e'-f'-g'-h'-i'$. Horizontal lines are now drawn through these points to meet vertical lines dropped from the points on the hyperbolic curve in the elevation. Points are thereby obtained through which the two hyperbolic curves a' to d' and f' to i', are drawn. The curves, one on each side of the middle section, are obtained as circular arcs using the point e' as centre. The two arcs join up with the hyperbolic curves, thus completing the contour of the joint line in the conveyor casing.

POINTS OR CURVES IN THE PATTERN

Cases often occur where, at certain points, the developed curve in the pattern should present a sharp point, while in apparently similar cases the sharp point should be curved or rounded off. These cases follow a definite rule which is illustrated in Fig. 68A.

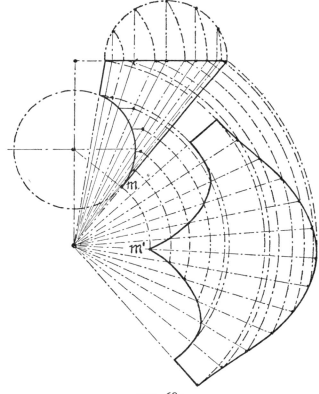

FIG. 68B

Three similar right cones are shown, at (*a*), (*b*) and (*c*). In (*a*) the right cone fits on a cylindrical surface such that the sides of the cone are tangential to the curve representing the cylinder, as seen at point *m*. This condition results in the curves meeting in sharp points, as at *m'* and *m"* in the pattern.

In (*b*), the right cone fits on a cylindrical surface such that the sides of the cone are cut through by the curve representing the cylinder, as seen at point *n*. This condition results in the curves in the pattern being rounded off, as at points *n'* and *n"*.

In (*c*), the right cone fits on an angular surface like a roof top, such that the sides of the cone are again cut through by the sides of the angular surface, as seen at point *o*. This condition again results in the curves in the pattern being definitely rounded, as seen at points *o'* and *o"*. Furthermore, since in this case the cone fits on an angular corner at *p*, the pattern presents a sharp corner at *p'* in somewhat the reverse order to the case at (*a*).

From these considerations it may be stated that where a cone or a cylinder is connected to or intersected by another geometrical body, and the sides of the cone or cylinder form tangents with the sides of the other body, then the curves in the pattern will form sharp corners at the points where the tangential contact occurs.

Alternatively, in cases where the sides of the cone or cylinder are cut through by the other body, then the curves in the pattern will be smooth and rounded at those points.

As a further example, the bottom side of the oblique conical hopper shown in Fig. 68B forms a tangent to the side of the cylindrical body at *m*. In consequence, the curves in the developed pattern form a sharp corner or point at *m'*.

4 Parallel Line Development

The parallel line method of pattern development depends on a process of locating the shape of the pattern on a series of parallel lines. All objects or articles which belong to the class of prisms, which preserve a constant shape of cross-section throughout their length, may be developed by the parallel line method. The general method of procedure is to "unroll" the surface. For instance, a cylinder is a prism with a circular cross-section at right angles to its central axis. An ordinary round pencil is a cylinder, and may be easily rolled along the table or any flat surface. An ordinary cylindrical pipe may be rolled along in the same way. From these examples it will be readily seen that the development of the pattern is equivalent to "unrolling" the surface.

A SQUARE RIGHT PRISM

Perhaps the simplest example which might be taken to illustrate this principle is an ordinary square prism. Fig. 69A shows a square right prism in plan and elevation. If this prism were rolled over from one side to another in the direction of the arrow, the pattern would be traced as shown at $A'B'C'D'A''$, and would be an ordinary

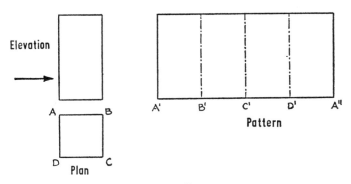

FIG. 69A

rectangle equal in length to the four sides of the square, AB, BC, CD and DA and equal in height to the height of the prism.

A SQUARE OBLIQUE PRISM

Consider now a similar prism, square at the base, but leaning over an angle to the plane of the base. An oblique prism such as this is shown in plan and elevation in Fig. 69B. To roll this prism over

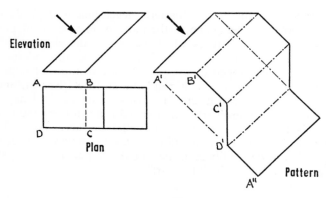

FIG. 69B

from one side to another, the direction of rolling would have to be at right angles to its axis, as indicated by the arrow. The body of the prism would then trace out the pattern as illustrated in Fig. 69B. In this case it should be noted that, in the pattern, the lengths of the sides of the square base, AB, BC, CD and DA, are equal to $A'B'$, $B'C'$, $C'D'$ and $D'A'$, but the actual widths of the panels forming the sides differ from one to another.

A HEXAGONAL RIGHT PRISM

As a further step in this method of development, the hexagonal right prism shown in Fig. 70A, is "unrolled" at right angles to its axis. The lengths of the six sides, AB, BC, CD, DE, EF and FA, are taken from the plan and marked off along the base line in the pattern, as from A' to A''. The height is, of course, equal to the height of the prism. Then the rectangle on $A'A''$ represents the pattern for the prism, with the bend lines shown at B', C', D', E' and F'.

Parallel Line Development

A HEXAGONAL OBLIQUE PRISM

The example shown in Fig. 70B represents a similar hexagonal prism leaning obliquely at an angle to the base. As in the case of the square oblique prism, the pattern is unrolled at right angles to its central axis, as indicated by the arrow. The full elevation is

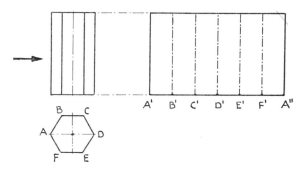

FIG. 70A

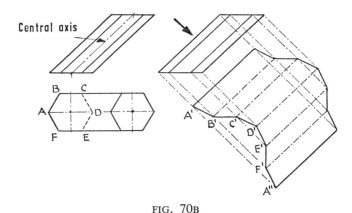

FIG. 70B

drawn, and lines projected from each of the points on the base and the top. Then, one of the equal sides of the hexagon such as AB is taken, and stepped over from one line to the next, as shown in the pattern at $A'B'C'D'E'F'A''$. This method ensures that the length of the zigzag line from A' to A'' is the same as the perimeter of the hexagon. From each of the points thus marked off from A' to A'', lines are now drawn parallel to the central axis of the prism, and, of course, parallel to each other. Where these parallel lines meet the lines projected from the top edge of the prism, points are afforded through which to draw the top edge in the pattern, which is similar in shape to the bottom.

A RIGHT CYLINDER

The development of the cylinder follows the same trend as that of the two previous examples. Fig. 71A represents a right cylinder in plan and elevation. There are no corners on this body which can

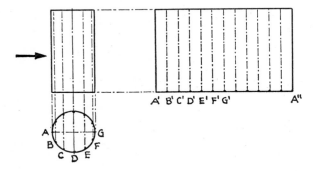

FIG. 71A

offer a natural means of dividing the perimeter into a number of parts. However, it is usual to divide the circular cross-section into twelve equal sections with the divisions spaced along the girth of the pattern. In Fig. 71A, the circle on the plan is divided into twelve equal parts and these are spaced along the base line of the pattern from A' to A''.

The length of $A'A''$ really ought to be equal to the circumference of the circle, but this method of taking the spacings from the circle straight across between the division points, instead of along the circumference, introduces a slight error in the total length. The difference amounts to 1·146 per cent, which, for most practical purposes, is unimportant. This difference is smaller still when the spacings are marked along a curved edge, as from A' to A'' in Fig. 71B.

Parallel Line Development

AN OBLIQUE CYLINDER

The example shown in Fig. 71B represents an oblique cylinder in which the base is a circle, but the central axis leans at an angle to the base. Any cross-section of the cylinder taken parallel to the base will be a circle, but any cross-section taken at right angles to the central axis will be an ellipse.

To develop the pattern, the circular base in the plan is divided into twelve equal parts, six of which are lettered from *A* to *G* in the Figure. The points on the circle are projected upwards to the base

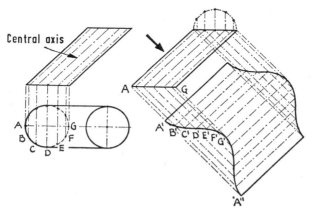

FIG. 71B

line in the elevation. From these points on the base, lines are drawn parallel to the central axis to the top extremity of the cylinder. The pattern is then "unrolled" from the elevation. However, in this case for the sake of convenient spacing, and a little modification of the process, the elevation is repeated on the right-hand side of Fig. 71B, and a semicircle is drawn on the top edge of the cylinder. The semicircle thus represents half of an inverted plan, and is divided into six equal parts. The points on the semicircle are dropped perpendicularly to the top edge, and from the points on the top edge lines are drawn parallel to the central axis to the bottom extremity of the cylinder.

Next, from the points on the top and bottom edges, lines are projected into the pattern at right angles to the central axis. This, of course, is the direction in which the pattern "unrolls." Then, one of the equal divisions is taken from the semicircle and, beginning at any point *A'* along the bottom line, this distance is stepped over from one line to the next and back again to *A"*. A curve drawn through those points gives the shape of the bottom edge in the pattern. To obtain the curve for the top edge, lines are drawn parallel to the central axis of the cylinder from each of the points on the curve $A'A''$. Where the parallel lines meet the corresponding lines projected from the top edge, points are afforded through which to draw the top curve in the pattern as shown in Fig. 71B.

Again, the distance along the curve from A' to A'' should be equal to the circumference of the circular base of the cylinder. In the present case the amount of error is somewhat less than in the previous example. This is because the line drawn from A' to A'' is a curve and not, as at in Fig. 71A a straight line.

AN OBLIQUE AND RIGHT CYLINDER Y-PIECE

Fig. 72 shows an oblique cylinder intersected by a right cylinder of equal circular section. Since the circular end section (not the

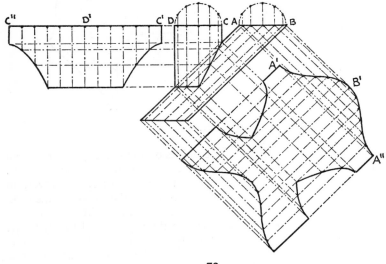

FIG. 72

cross-section) of the oblique cylinder is of the same diameter as the right cylinder, the shape of the joint line between the two cylinders is represented in the elevation by two straight lines meeting at the centre.

The development of the pattern for the oblique cylinder is similar to that of the example in Fig. 71B, with the addition of the shape of the hole or joint line. To determine the shape of the hole in the pattern,

Parallel Line Development

the points on the joint line in the elevation are projected into the pattern at right angles to the central axis of the cylinder. Where these lines meet or cross the corresponding lines in the pattern, points are afforded through which to draw the joint curve, as shown in the Figure.

The pattern for the right cylinder is unrolled at right angles to its central axis, and the spacings from the semicircle are marked off along the top edge of the pattern. The plotting of the shape of the joint edge should be self-evident from the illustration.

It should be noted that the length of the circular edge AB is spaced along the curve $A'B'A''$, and the length of the circular edge CD is spaced along the straight line $C'D'C''$.

A RIGHT CYLINDRICAL CHUTE

Fig. 73 shows a delivery chute from a screw conveyor. The chute is of right cylindrical construction, and the joint line at the elbow bisects the angle of diversion.

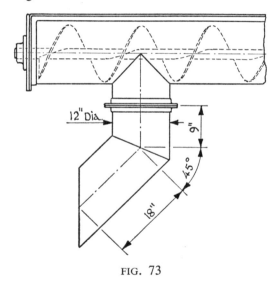

FIG. 73

The development of the patterns for the chute is shown in Fig. 74. As both parts of the chute are portions of right cylinders, the diameters and circumferences at right angles to the central axes are the same for both. A semicircle is drawn on the top edge of the chute, divided into six equal parts and the points numbered from 1

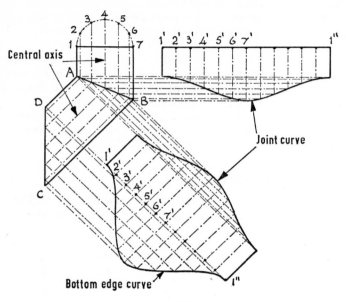

FIG. 74

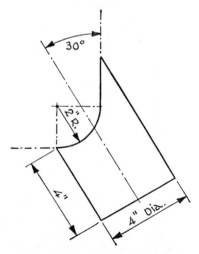

FIG. 75. *Outlet connexion from tank bottom*

Parallel Line Development

to 7. From the points on the semicircle, lines are dropped perpendicular to the top edge, and continued to the joint line AB. From the points thus obtained on AB, lines are drawn parallel to the central axis to meet the bottom edge CD of the chute.

The top part of the chute is unrolled horizontally, and the divisions from the semicircle are spaced along the line $1'7'1''$. From the points on $1'1''$, parallel lines are drawn perpendicularly to meet horizontal lines projected from the joint line AB. Thus points are afforded through which to draw the joint curve.

The bottom part of the chute is unrolled at right angles to its central axis, and as this is also a portion of a right cylinder, the girth line or circumference of the cross-section must be spaced along one of the straight lines projected from the elevation. To obtain the pattern for this section, lines are projected from the points on the joint line AB and also from the points on the bottom edge CD, all at right angles to the central axis. The divisions are taken from the semicircle, and, in this case, are spaced along the line projected from point D, as shown from $1'$ to $1''$. Through all the points between $1'$ and $1''$, lines are drawn parallel to the central axis. Where these lines cross the lines projected from the elevation, points are afforded through which to draw the joint curve and bottom edge curve as shown in Fig. 74.

OUTLET CONNEXION TO TANK BOTTOM

The example shown in Fig. 75 represents a cylindrical outlet connexion to a tank with a rounded corner at the bottom. In this case the corner radius is 2 in. and the diameter of the cylindrical connexion 4 in., and the connexion fits at an angle of 30° to the vertical.

The development of the pattern is shown in Fig. 76, in which the pattern is unrolled at right angles to the central axis of the pipe. A semicircle is first drawn on the base of the pipe and divided into six equal parts. Lines are then drawn parallel to the centreline from the points on the semicircle straight through to meet the joint line at the top. The base line is next extended into the pattern, and twelve spaces equal to those around the semicircle are marked off along the base line. From the points on the base line and at right angles to it, a series of parallel lines are drawn to meet corresponding lines projected parallel to the base line from the points on the joint line in the elevation. Where the points projected from the joint line meet the parallel lines erected from the base line in the pattern, points are afforded through which to draw the joint curve in the pattern.

74   *Sheet Metal Drawing and Pattern Development*

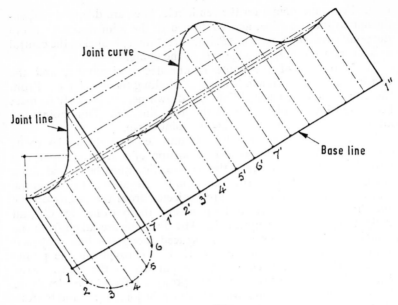

FIG. 76

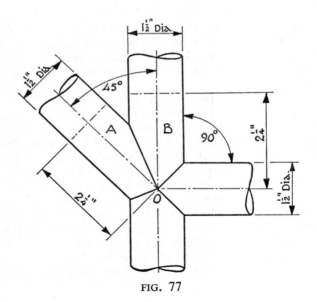

FIG. 77

Parallel Line Development

CROSS PIPES OF EQUAL DIAMETER

Fig. 77 shows four pipes of equal diameter meeting at a single point *O*. Since the pipes are of equal diameter the joints between them are represented by straight lines as shown in the Figure. The patterns for the two parts *A* and *B* are shown developed in Fig. 78. The procedure is similar to that followed in the previous examples of right cylindrical pipe development, inasmuch as semicircles are drawn on the ends or cross-sections of the pipes as shown in the

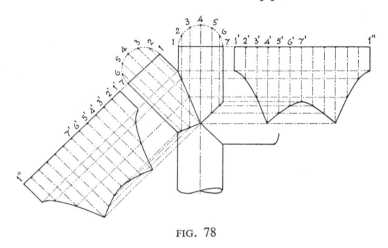

FIG. 78

Figure. The semicircles are divided into the usual six equal parts and lines projected parallel to the centrelines from the points on the semicircles to the joint lines which meet at the centrepoint *O*. The patterns are then unrolled at right angles to the respective centrelines as shown in Fig. 78. The remainder of the procedure should now be self-evident from the illustration.

A SIMPLE MOULDING

Fig. 79 represents a simple moulding which could be used as a curb, as shown, or a gutter if turned upside down. To develop the pattern the cross-section at right angles to its central axis is divided into a convenient number of parts and numbered, as shown in the Figure, from 1 to 10. From the points on the cross-section, lines are dropped vertically to meet the joint line *AB* in the plan. Then, from the top end of the section, and from the points on *AB*, lines are projected horizontally into the pattern. Next, the divisions are taken in

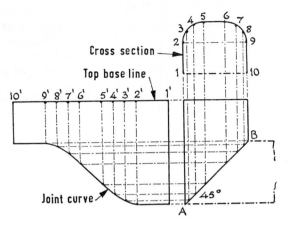

FIG. 79

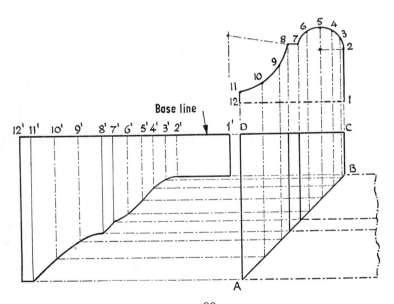

FIG. 80

Parallel Line Development

consecutive order from 1 to 10 on the cross-section, and spaced along the top base line, as shown from 1' to 10'. Now, from the points on the top base line, vertical lines are drawn to meet the corresponding horizontal lines from the joint line AB. Points are thereby afforded through which to draw the joint curve in the pattern.

A CURB MOULDING

Fig. 80 also represents a type of moulding suitable for a curb. The method of developing the pattern is precisely the same as that of the moulding in Fig. 79, in so far as the cross-section is divided

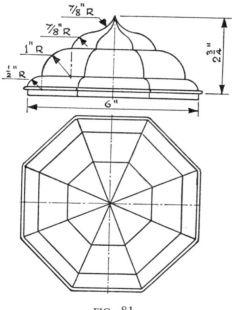

FIG. 81

into a convenient number of parts and the points numbered, though they are numbered from 1 to 12 in this particular case. Lines are then dropped vertically from the points on the cross-section to meet the joint line AB in the plan. Then, the section $ABCD$ is unrolled at right angles to its axis, or parallel to its top edge CD. Thus, CD is extended into the pattern, and the divisions from the cross-section are marked off from 1' to 12' along the top base line. Vertical lines are now dropped from the points on the base line to meet horizontal lines projected from the points on the joint line *AB*. Where the verticals meet the corresponding horizontals, points are afforded through which to draw the joint curve in the pattern.

A SEGMENTAL LID

The octagonal lid shown in Fig. 81 is a further typical example of parallel line development. Segmental work of this kind may have

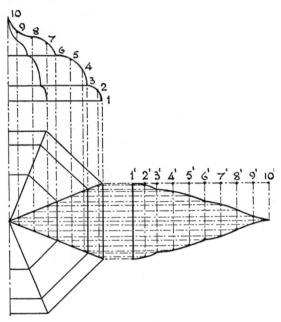

FIG. 82

a square, hexagonal, octagonal, or any other polygonal cross-section.

The development of the pattern for one segment is shown in Fig. 82. Since the centreline of the segment chosen is horizontal in the plan, the shape of the outline in the elevation represents the true contour of the centreline. To develop the pattern, the outline in the elevation is divided into a convenient number of parts, and numbered as shown in the Figure from 1 to 10. Then, from these points, lines are dropped vertically into the plan to cross the segment from side to side.

Parallel Line Development

Next, the centreline of the segment in the plan is extended to serve as the centreline in the pattern. The divisions on the outline in the elevation are now taken and marked off along the centreline in the pattern. Alternatively, these divisions may be spaced along a parallel line on the outside of the pattern, as shown in Fig. 82. This is, of course, equivalent to spacing them along the centreline. From the points thus marked off, lines are drawn across and at right angles to the centreline. Next, lines are projected parallel to the centreline from all the points on the two sides of the segment to

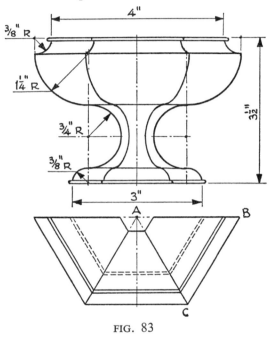

FIG. 83

meet the lines drawn across the centreline. Where the lines parallel to the centreline meet the corresponding perpendiculars, points are afforded through which the two side contours of the pattern are drawn as shown in Fig. 82.

As all the segments are the same, one pattern serves for marking off the remainder.

A HEXAGONAL VASE

The illustration of a hexagonal vase shown in Fig. 83 is typical of the way in which many of these examples are presented. The plan shows half the hexagon, with the horizontal centreline of the hexagon along two opposite corners. This means that the shape of the outline in the elevation is not the true contour of the centreline of one segment, but represents the contour of a corner.

There are two methods by which the pattern for one segment can be obtained when the elevation presents a corner contour. One method is to project a true contour of the centreline of the segment and step the divisions along the centreline in the pattern, as in Fig. 84. The other method is to take the divisions direct from the corner contour in the elevation and step them over from one line to another along the parallel lines on each side of the pattern after the manner of oblique cylinder development, as in Fig. 86. This method ensures that the length of the curved edge in the pattern is the same as that of the corner contour in the elevation.

The method of development shown in Fig. 84 is that of projecting a true contour of the centreline. Only one segment, ABC, in the plan is needed. The corner contour in the elevation is divided into a convenient number of parts, and lines are drawn vertically downwards from all the points on the contour to the corner edge AB in the plan. From the points on AB the lines are continued, but diverted in a direction parallel to BC to meet the other corner edge AC. The centreline of the segment is represented by AD. The point D is, of course, the middle point of BC, and AD is at right angles to BC.

Now, a true contour of the centreline will be seen at right angles to AD in the direction of the arrow. Therefore, all the lines on the segment parallel to BC are projected forward, and a new base line is drawn across them at right angles in any convenient position.

Before the next step is explained, it is important to observe that the vertical height of the vase, $3\frac{1}{2}$ in., will be the same in the projected view as it is in the elevation. From this it follows that the vertical heights from the base line of each and all the points on the corner contour in the elevation will be the same in the projected view. Therefore, the next step is to take the vertical height of each point from the base line in the elevation, and mark that height on the corresponding line from the new base line in the projected view.

For example, the vertical heights of the two points 6 and 12 in the elevation, both on the same vertical line from point b on the base line, are taken and marked upwards on the projected line from b'' in the true contour to give points $6'$ and $12'$. The point b'' may readily be located by following the vertical line downwards from b to b', and then continuing up the projected line to b''. This process repeated with the other points on the corner contour will provide sufficient points on the true contour of the centreline to draw the curve as shown in Fig. 84.

Parallel Line Development

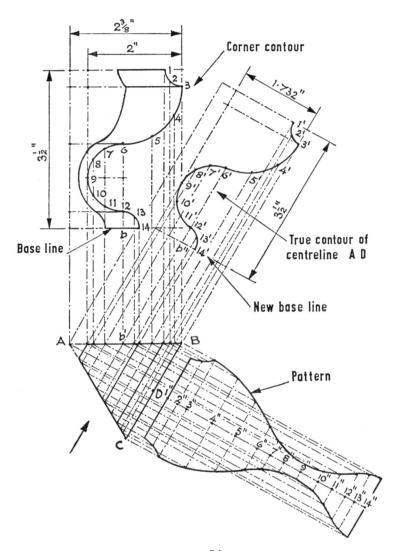

FIG. 84

The form of the true contour curve in the projection is very similar to that of the corner contour in the elevation. The difference lies in the horizontal width, which is 2 in. in the elevation but only 1·732 in. in the projected view. This means that the actual length of the curve in the true contour is shorter than that of the corner contour.

It remains now to develop the pattern. The centreline AD in the plan is extended to serve as the centreline in the pattern. Now, as the length of the curve in the projected view is the true length of the centreline AD, the spacings on the curve from 1' to 14' are taken and marked off along the extended centreline, as from 1" to 14". Through these points, lines are drawn across and at right angles to the centreline. Next, from all the points on the two corner edges AB and AC in the plan, lines are projected parallel to the centreline into the pattern. Where these lines meet or cross the lines at right angles to the centreline, points are afforded through which to draw the outside curves in the pattern as shown in Fig. 84.

A SIMPLE FINIAL

The example shown in Fig. 85 represents a simple type of finial or ornamental top to a spire or gable. There is infinite scope for

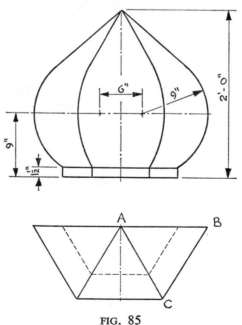

FIG. 85

Parallel Line Development

variety in the shape and design of finials, and all those of segmental construction lend themselves to parallel line development. The present example is one of hexagonal design, and the plan view presents half of the hexagon with the corner edges in the horizontal position. This means that the outside contour in the elevation is a corner contour and not a contour of the centreline of a segment.

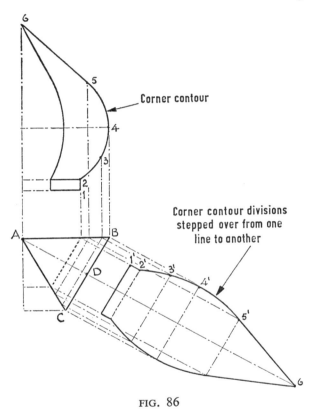

FIG. 86

The method of development adopted in this example, Fig. 86, is the alternative to that used in Fig. 84. In this case the corner contour in the elevation is again divided into a convenient number of parts, as shown in the Figure from 1 to 6. From the points on the contour, lines are dropped vertically to the corner edge AB in the plan. From the points obtained on the corner edge, lines are drawn parallel to BC to meet the opposite corner edge AC. Next, the centreline AD is extended to serve as the centreline of the pattern.

Then from the points on the two corner edges AB and AC, lines are projected into the pattern parallel to the centreline.

The next step embodies the difference between the present method and that of the previous example. In this case a projected contour of the centreline AD is not required for spacing along the centreline of the pattern. Instead, the divisions are taken from the corner contour in the elevation, as from 1 to 6, and, beginning in any convenient position on the line obtained from point 1, the divisions are stepped over from one line to another, as shown from 1' to 6'. It is important to observe that the divisions must be stepped on to the correct line corresponding to that obtained from the elevation, not necessarily the next adjacent line. For example, it will be seen in the Figure that from point 2' to point 3' the division steps over the line which goes to point 5'. This method ensures that the length of the edge curve in the pattern from point 1' to 6' is the same as that of the corner contour in the elevation. A similar curve on the other side of the centreline completes the pattern for one segment, as shown in Fig. 86.

AN ORNAMENTAL FLOWER STAND

As the flower stand shown in Fig. 87 is octagonal, and the horizontal centreline in the plan coincides with the centreline of one segment, the outside contour in the elevation represents the true contour of the centreline of a segment.

The development of the pattern is shown in Fig. 88. In the plan, one segment is placed with the centreline in the horizontal position. The contour in the elevation is divided into a convenient number of parts and the points numbered as in the illustration from 1 to 15. The points on the contour are dropped into the plan to cross the segment ABC from side to side.

The centreline of the segment ABC is now extended to serve as the centreline in the pattern, and all the points on the two sides AB and AC are also projected parallel to the centreline into the pattern.

In Fig. 88 it will be seen that all the lines projected parallel to the centreline from the two edges AB and AC in the plan are arrested on the line $B'C'$ and, with centre O, revolved through 90° to the position $B''C''$, O being any suitable point on the line $C'B'$ extended. All the lines are then continued vertically from $B''C''$. The reason for this manœuvre is to bring the pattern into a more convenient position for compactness of space, since the projection of the pattern horizontally would have resulted in an unduly wide

Parallel Line Development

illustration. This method may often be used to displace a pattern to a more convenient position.

The remainder of the process is similar to that described in connexion with Fig. 82, in so far as the divisions are taken from the contour in the elevation and spaced along the centreline of the pattern. Lines are then drawn through those points at right angles

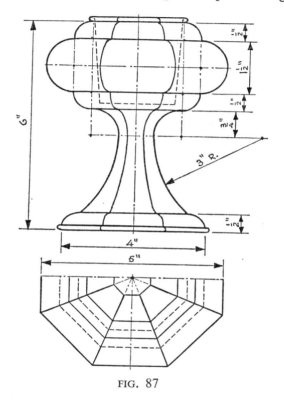

FIG. 87

to the centreline, and where they meet the corresponding lines projected round the Figure from the contour in the elevation, points are afforded through which to draw the outline of the pattern.

AN ELONGATED SEGMENTAL BOWL

In the previous examples of segmental work, the shape of the object in the plan was either a regular hexagon or regular octagon, which gave equal and similar segments all round. In the elongated

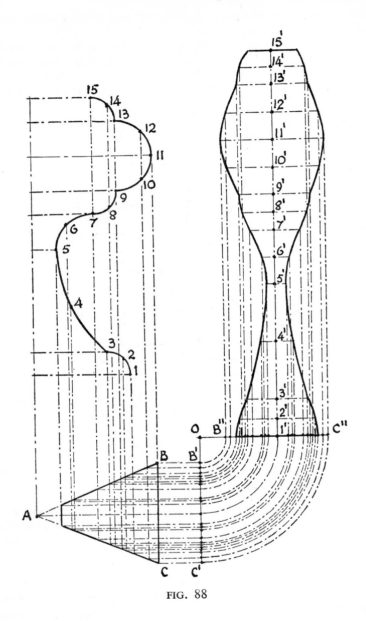

FIG. 88

Parallel Line Development 87

octagonal bowl shown in Fig. 89, three different patterns are required for the segments *A*, *B* and *C*. The remaining segments would, of course, be duplicates of these three. The development of the patterns is shown in Fig. 90. Only one quarter of the plan is used, showing half of the segment *A*, a full segment *B*, and a half segment *C*.

As the centreline of segment *A* in the plan is in a horizontal position, the outside contour in the elevation is the true contour

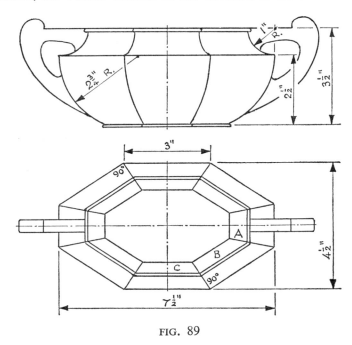

FIG. 89

of the centreline of the segment. The contour in the elevation is divided into a convenient number of parts, as shown in the Figure from 1 to 5. Lines are dropped from these points to cross the half segment *A* and meet the joint line with segment *B* in the plan, and are then continued onward parallel to the axis of segment *B* to meet the joint line with segment *C*. The centreline of *A* is extended to serve as the centreline in the pattern, and the divisions are taken from the contour in the elevation and marked off along the centreline in the pattern as shown from 1' to 5'. Lines are now drawn through these points at right angles to the centreline to meet corresponding lines projected horizontally from the joint line between the segments *A* and *B*. Points are thus afforded through which to draw the joint curve in the pattern. The pattern for segment *A* is completed by drawing a similar but opposite curve on the other side of the centreline as shown in the Figure.

The pattern for segment *B* is next unrolled at right angles to its axis. The points on the joint line between segments *A* and *B* are projected into the pattern at 90° to the outside edge of segment *B*.

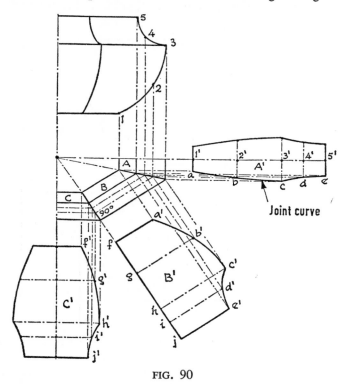

FIG. 90

As the other joint line with segment *C* is at 90° to the outside edge of segment *B*, that edge is projected into the pattern as a single straight line, as is shown in the Figure. The next step should be carefully observed. The spacings along the curve in the pattern for segment *A*, as at *ab*, *bc*, *cd* and *de*, are taken. Then, beginning at any point *a'* on the corresponding line projected from the joint line into pattern *B*, these spacings are stepped over from one line to another. It is important to observe that each line marked off should correspond to that projected from the same point on the joint line into each pattern. This method ensures that the length of

Parallel Line Development

the joint curve in the pattern A' is the same as the length of the corresponding joint curve in pattern B'. From the points a', b', c', d' and e', lines are drawn across the pattern at right angles to those projected from the joint line to meet the single line on the other side.

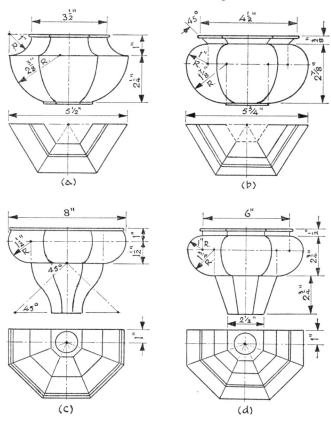

FIG. 91

The spacings obtained on the single line represent the length of the joint line between segments B and C.

The pattern for segment C may now be obtained. The centreline of the segment is extended to serve as the centreline in the pattern. Then all the points on the joint line between segment B and segment C are projected parallel to the centreline into the pattern. Now, the spacings f–g–h–i–j on the straight joint line of pattern B', represent the actual length of the joint between the two segments B and C.

This length must be the length of the curved joint edge of pattern C'. Therefore, the spacings *fg*, *gh*, *hi* and *ij* are taken from pattern B' and, beginning at point f' on the inside line of pattern C', these spacings are stepped over from one line to another, care being

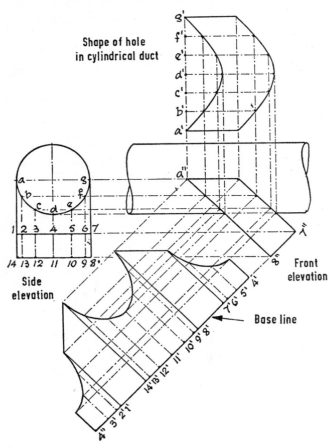

FIG. 92

taken to see that the correct line is marked in the process of spacing. As in the case of pattern A', that shown at C' is only one half of the complete pattern. The other half is drawn symmetrically opposite on the other side of the centreline.

Further examples for practice in the development of segmental work are given in Fig. 91.

Parallel Line Development

AN OBLIQUE RECTANGULAR BRANCH ON A CYLINDRICAL BODY

Fig. 92 represents a rectangular pipe intersecting a cylindrical pipe at an oblique angle. The development of the pattern for the branch pipe and the shape of the hole in the cylindrical pipe are obtained by the parallel line method. The width of the rectangular pipe is equal to the diameter of the cylindrical duct.

For the convenience of development, the bottom half of the circle in the side elevation is divided into six equal parts and lines from those points are drawn vertically downwards on the surface of the rectangular pipe. Also, from the points on the semicircle, lines are drawn horizontally to meet the top and bottom edges of the rectangular pipe in the front elevation. The pattern is now unrolled at right angles to the axis of the rectangular duct.

From the bottom edge of the rectangle in the front elevation the base line is projected into the pattern. Next, the divisions are taken from the rectangle in the side elevation, beginning at the top centre point 4 and working round the rectangle in a clockwise direction; these spacings are then marked off along the base line. One exception must be observed. The distances between points 7 and 8, and also between 1 and 14 in the side elevation are not true distances. These divisions are taken from the front elevation from $7''$ to $8''$. From the points marked along the base line, lines are drawn at right angles to meet the series of parallel lines projected from the two edges in the front elevation. Points are thereby afforded through which to draw the shape of the top edge in the pattern as shown in the Figure.

It now remains to develop the shape of the hole in the cylindrical pipe. The hole extends half-way round the pipe as from a to g in the side elevation. Therefore, at right angles to the central axis of the duct, the half circumference is projected upwards from point a'' and the spacings from a' to g' are made equal to those in the side elevation from a to g. Then, from points a', b', c', d', e', f' and g', lines are drawn horizontally to meet vertical lines drawn from the corresponding points on the two adjacent lines in the front elevation. This process should be readily followed from the illustration. Points are thereby afforded through which to draw the shape of the hole as shown in Fig. 92.

AN ELONGATED OBLIQUE BRANCH ON A CYLINDRICAL BODY

The example given in Fig. 93 is similar to that of Fig. 91, but with the difference that the rectangular branch pipe has semicircular ends. This may best be seen in the side elevation, though the semi-circles appear as semi-ellipses in that view.

To develop the pattern, a semicircle is first drawn on the lower end of the branch in the front elevation and divided into the usual

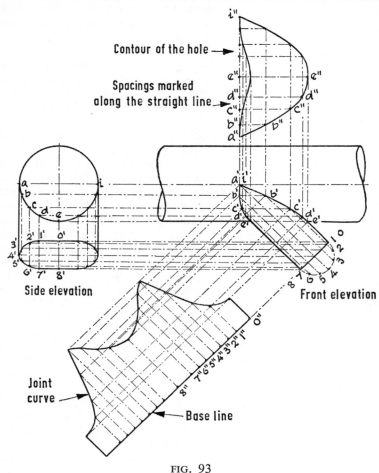

FIG. 93

six equal parts. The points are numbered from 1 to 7, as shown in the illustration, and projected back perpendicular to the end of the branch. From the points on the end of the branch, lines are projected horizontally into the side elevation and the semi-ellipses drawn at each side of the branch in that view. The points on the left-hand semi-ellipse are numbered from 1' to 7', and also the straight

Parallel Line Development

portions back to the centreline from 1' to 0' and 7' to 8'. The points on the semi-ellipses are also projected upwards to the lower semi-circle in the side elevation, and lettered a, b, c, d and e.

Now, to obtain the shape of the intersection in the front elevation, the lines projected from the points 1 to 7 to the end of the branch are continued onward parallel to the centreline. Then, from the points a, b, c, d and e in the side elevation, horizontal lines are drawn to meet the corresponding lines drawn to the branch in the front elevation. Thus, points a', b', c', d' and e' are located on each side of the centreline as shown in the front elevation. Also, the joint line extends round the opposite side of the cylindrical pipe to point i, and in the front elevation lies directly behind the joint line as seen from a' to e'.

To unroll the pattern, the base line from the front elevation is extended, and, beginning at any convenient point 0'', the distance 0''1'' is marked off equal to 0'1' in the side elevation. Then, the divisions from 1 to 7 are taken from the semicircle in the front elevation and marked off along the base line in the pattern from 1'' to 7''. Next, the distance from 7' to 8' is taken from the side elevation and marked off along the base line from 7'' to 8''. All the spacings thus marked off from 0'' to 8'' are now repeated in the reverse order along the base line in the pattern. From these points, lines are now drawn at right angles to the base line to meet corresponding lines projected from the joint line in the front elevation. For example, the line in the pattern drawn from point 5'' meets the line projected from b' on the joint line which in turn is obtained from point 5 on the semicircle. In this way points are obtained through which to draw the joint curve in the pattern as shown in Fig. 93.

The contour of the hole in the cylindrical pipe may now be obtained by unrolling at right angles to the central axis of the pipe. Thus, lines are drawn vertically upwards from all the points on the joint line in the front elevation. Since the hole extends half-way round the pipe, as from a to i in the side elevation, the spacings are taken from the half circle from a to i and marked off along the line projected from a' in the front elevation, as shown from a'' to i''. Horizontal lines are now drawn from these points to meet the lines drawn vertically from the joint line in the elevation. Points are thereby afforded through which to draw the shape of the hole as shown in Fig. 93.

CYLINDRICAL PIPE TO SQUARE PYRAMID

Fig. 94 represents a square-based pyramid intersected by a cylindrical uptake pipe or duct. It is necessary, before the patterns can be developed, to determine the form and position of the joint line. Each panel of the pyramid intersects the cylinder in a plane at an angle to the axis of the cylinder. Therefore, the shape of the curve at each section of the joint line is elliptical.

The determination of the joint line and development of the patterns are shown in Fig. 95. First, a half plan is drawn on the base of

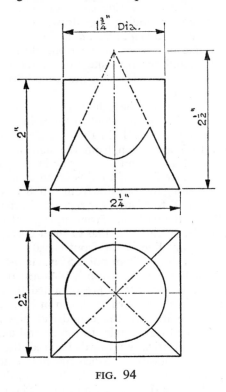

FIG. 94

the elevation, and the front panel of the pyramid is lettered *ABC*. The base *BC* of the panel is divided into four equal parts, as at *a*, *b* and *c*, and the points are joined to the apex *A*. The points *a*, *b* and *c* are next projected perpendicularly to the base *B'C'* in the elevation and the points of intersection are lettered *a'*, *b'* and *c'*. From these points lines are drawn to the apex *A'* in the elevation.

The joint curve *x'z'w'* in the elevation is obtained from the corresponding points in the plan. Thus, the corner line *BA* crosses the semicircle representing the cylinder at the point *x*. This point *x* is now projected vertically upwards to the corner line *B'A'* in the

Parallel Line Development

elevation to obtain the point x'. Similarly, the plan line aA crosses the semicircle in point y, and the point y is projected vertically upwards to the corresponding line $a'A'$ in the elevation to obtain the point y'. Now, the point z' in the elevation is at the same vertical height above the base as the point V' where the outside of the cylinder cuts the outside line $B'A'$ in the elevation. Therefore, point V' is

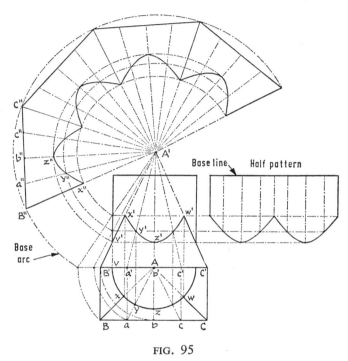

FIG. 95

projected horizontally to the centreline to obtain the point z'. The curve $x'y'z'$ is repeated on the other side of the centreline to point w'.

To develop the pattern for the pyramid, the plan lengths AB, Aa and Ab are swung round to the base produced, and true length lines are drawn from those points to the apex A'. Next, points x', y' and z' are projected horizontally to the corresponding lines just drawn. Then, using the apex A' as centre, the points on the true length lines are swung into the pattern, including the outside base arc. Now, the length BC is taken from the plan and marked off four times around the base arc in the pattern. Straight lines are then drawn between the points on the arc, and each line divided into four equal parts corresponding to those on BC in the plan. Thus, one section in the pattern is lettered $B''a''b''c''C''$. Radial lines are now drawn to the apex A' from all the points around the base. Where the radial lines cross the arcs from the elevation, points are afforded through which to draw the joint curves in the pattern.

To develop the pattern for the cylinder, the base line at the top is extended, and divisions marked off along the extension equal to those on the quadrant in the plan from x to w. This, of course, is one quarter of the full circumference, and would need to be repeated four times for the full pattern. Parallel lines are drawn perpendicular to the base line from all the points marked on it. Then, from the points x', y' and z' in the elevation, lines are drawn parallel to the base line to meet or cut the perpendicular lines, thereby affording points through which to draw the joint curves in the pattern as shown in Fig. 95.

5 The Common Central Sphere

The type of intersection of right cones and cylinders which presupposes the existence of a common central sphere is of particular importance in sheet metal pattern development. The chief advantage is that in a front elevation of the intersecting bodies, the shape of the joint presents a straight line. Leading examples of this type of intersection are given in Fig. 96. At (*a*), Fig. 96, the circle represents a sphere which fits inside the right cone *A*, thereby making the sides of the cone tangential to the circle. The cylinder *B* inclines at an angle to the axis of the cone *A*, such that the central axis of the cylinder passes through the centre of the sphere. The diameter of the cylinder is equal to that of the sphere, which make the sides of the cylinder tangential to the circle. Now, the joint line between the cone and the cylinder is represented by the straight line from *a* to *b*. The point *a* is situated where the inner line of the cone crosses the inner line of the cylinder, and the point *b* is where the outer line of the cone meets the outer line of the cylinder. It should be especially noted that the joint line *ab* does *not* pass through the centre of the circle.

At (*b*), Fig. 96, the circle again represents a sphere which fits inside the right cone *A*, thereby making the sides of the cone tangential to the circle. Another cone *C* inclines at an angle to the axis of the cone *A*, such that the central axis of the cone *C* passes through the centre of the sphere. The two sides of the cone *C* are drawn tangentially to the circle. Now, the joint line between the cone *A* and the cone *C* is represented by the straight line from *c* to *d*. The point *c* lies where the inner line of cone *A* crosses the inner line of cone *C*, and the point *d* lies where the outer line of cone *A* meets the outer line of cone *C*. Again, it should be particularly noted that the joint line *cd* does *not* pass through the centre of the circle.

Fig. 96c shows the combined intersection of *A*, *B* and *C*. The joint line between *A* and *B* is shown at *ab*, and the joint line between *A* and *C* is shown at *cd*. In addition, this example involves a joint line between *B* and *C*, and if these two bodies be considered alone, the joint line would occur between *e* and *f*. It is important to note that the three joint lines intersect at a single point *p*, which again, it should be noted, is *not* the centre of the circle. In this compound intersection of three bodies, only portions of the separate joint lines are involved in the actual construction. Between *A* and *B* the portion *pa* is required; between *B* and *C* the portion *pf* is needed; and

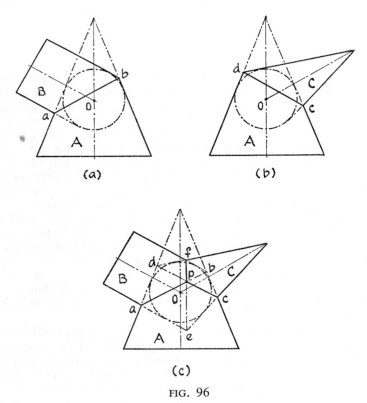

FIG. 96

between *C* and *A* the portion *pc* completes the pattern of the joint lines.

Thus, the three examples in Fig. 96 illustrate the basic principle of intersections around a common central sphere, from which two important points arise. First, the central axes or the centrelines of the intersecting bodies all pass through the centre of the sphere, or the circle representing the sphere. Second, the joint lines do *not* pass through the centre of the circle representing the sphere, and in cases of more than two intersecting bodies, the several joint lines

The Common Central Sphere

cross at a single point which, again, does not lie at the centre of the circle.

TWO INTERSECTING CONES

The expanding nozzle shown in Fig. 97 is composed of two right cones with central axes intersecting at right angles. The joint line is represented by the straight line *AB*, and since there is only one condition on which this depends, the inference is that the two cones intersect around a common central sphere. The centre of the circle

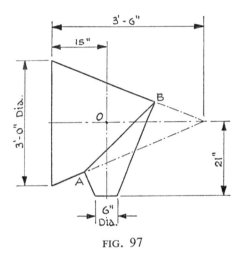

FIG. 97

representing the sphere will occur at *O* where the two centrelines cross.

The solution of the problem and the development of the patterns is shown in Fig. 98. First, the larger cone is drawn with its axis horizontal, and the centrepoint *O* is located through which the vertical axis of the smaller cone is drawn. With centre *O*, the circle representing the sphere is then drawn so that the circle just touches the sides of the cone, thus making the two sides tangents to the circle. The position of the smaller vertical cone is now located by drawing the sides of the cone from the 6 in. diameter as tangents to the circle. Then, where the two inside lines of the cones meet or cross, the point *A* is obtained, and where the two outside lines of the cones meet, the point *B* is obtained. The straight line *AB* then represents the joint line.

The patterns are obtained by straightforward right conic radial line development. In the case of the larger horizontal cone, a semi-circle is described on the base *CD*, divided into six equal parts, and lines drawn from the points back to the base. From the points on the base, lines are then drawn to the apex A'. Where these lines cross the joint line *AB*, lines are drawn from the points thereon to the side of the cone at right angles to the centreline. Next, with the

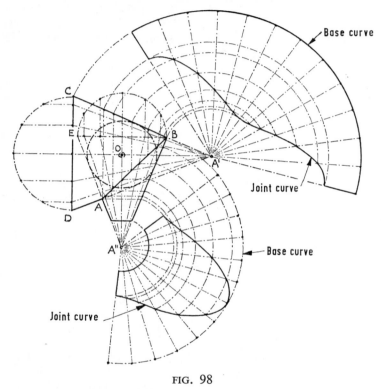

FIG. 98

apex A' as centre, all the points on the side $A'C$ of the cone are swung into the pattern, and twelve spaces, equal to those on the semicircle, are marked off along the outside arc to form the base curve in the pattern. From the points on the base curve, radial lines are drawn to the apex A'. Where the radial lines cross the arcs, points are afforded through which to draw the joint curve in the pattern.

To develop the pattern for the smaller cone, a base line *EB* is first drawn at right angles to the centreline of the cone. The process of development is then similar to that of the larger cone. A semicircle is described on the base EB, divided into six equal parts, and the points dropped perpendicularly to the base line. From the points on the base, lines are drawn to the apex A'', and from the points where these lines cross the joint line AB, lines are drawn to the side $A''B$ at right angles to the centreline of the cone. Then, using the apex A'' as centre, all the points on the side $A''B$ are swung into the pattern. Twelve spaces are marked off along the outer arc or base curve, and from the points thereon radial lines are drawn to the apex A''. Where the radial lines cross the remaining arcs, points are afforded through which to draw the joint curve in the pattern as shown in Fig. 98.

A TWO-POINT RIGHT CONICAL NOZZLE

The example shown in Fig. 99 represents a further application of the principle of the common central sphere. The two horizontal nozzles are of right conic construction intersected by the cylindrical

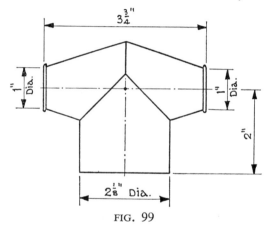

FIG. 99

pipe. Since the two nozzles are similar, only one, with the cylindrical pipe, is shown developed in Fig. 100.

In setting out the elevation from the given measurements, the two centrelines are first drawn, and the circle representing the sphere constructed with its centre at O where the two centrelines cross. The diameter of the circle must, of course, be equal to the diameter of the cylinder, which is next drawn with its sides as tangents to the circle. Next, the position of the smaller diameter of the conic nozzle is located, and from the extremities of that diameter tangents to the circle are drawn. The conic tangents are produced backwards from the diameter to obtain the apex A of the cone. Now, assuming that the nozzle consisted of the cylinder and one cone only, as shown, then the joint line between the cylinder and the cone would occur as from B to C where the inside and outside lines of the two bodies meet. However, the joint line which would occur between the two cones would be a straight line between their bases and would coincide

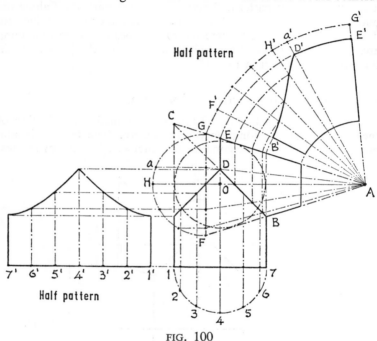

FIG. 100

with or fall on the centreline of the cylinder. From this it will be seen that the portion of the joint line required between the cones will be that from D to E, and the full joint line needed for the development of the cone will be as from B to D to E.

The first step in developing the pattern for the cone is to assume a position for the base. In Fig. 100 the base line is drawn at FG. It might be well at this stage to note that a base line could be taken through the centre of the circle by extending ED to the other side of the cone. This has the disadvantage that the radius of the base, OE, would be only slightly larger than the radius of the circle representing the sphere, and this often leads to confusion. Therefore, as the base for the cone may be placed in any convenient position,

The Common Central Sphere

it is wise to push it back sufficiently far from the centre to avoid any possibility of mistake.

With the base assumed at *FG*, the usual semicircle is drawn on it and divided into six equal parts, though only three of these parts on the quadrant from *F* to *H* will be needed. The points on the quadrant *FH* are projected back to the base line, and from the points on the base line, lines are drawn to the apex *A*. From the points where these lines cross the joint line *BD*, lines are drawn at right angles to the centreline to the outside, *AG*, of the cone. Then, from all the points on *AG*, arcs are swung into the pattern, and the three spaces from the quadrant *FH* are marked off along the outside arc as from *F'* to *H'*. For the half pattern a further distance from *H'* to *G'* is marked off equal to *F'H'*. It still remains to mark off the special point *a'*. To determine this point, a line is drawn from the apex *A* through the point *D* to the base line of the cone. Then the point obtained on the base line is projected parallel to the centreline to locate the point *a* on the semicircle. The distance *Ha* is now taken from the semicircle and marked off from *H'* along the outside arc, as from *H'* to *a'*. Radial lines are now drawn from the points on the outside arc to apex *A*, and where the radial lines meet the remaining arcs, points are afforded through which to draw the base curve in the pattern, as shown in Fig. 100.

In Fig. 100, one half of the pattern for the cylinder is shown unrolled on the left of the elevation. This is obtained by straightforward parallel line development, in which the base 1,7 is extended into the pattern, and the divisions from the semicircle marked off along the base line in the pattern. From the examples of cylinder development previously dealt with, the remainder of the process should not be difficult to follow from the illustration.

A RIGHT CONICAL OUTLET FROM A CYLINDRICAL DUCT

The right conical connexion to the cylindrical duct shown in Fig. 101 represents another example of intersection around a common central sphere. The solution of the problem and the development of the pattern for the cone are shown in Figs. 102 and 103.

First, the right cone is drawn to the given measurements and then the centre of the circle which represents the sphere is located. The diameter of the circle will be the same as the diameter of the cylinder, and to locate its centre on the centreline of the cone, the radius of the circle is taken in the compasses and two arcs drawn from any two points on the side of the cone. A straight line drawn as a tangent to the arcs will pass through the required centre on the centreline of the cone, as shown in Fig. 102. The straight line need not be drawn in full. All that is necessary is a short dash across the centreline of the cone to locate the centre of the sphere. The circle may then be drawn, which clearly should just touch the sides of the cone. The cylinder may next be drawn, with its sides tangential to the circle.

Now, the joint line between the cone and the cylinder is in two parts, from *B* to *D* and from *D* to *E*. The part *BD* is a portion of the

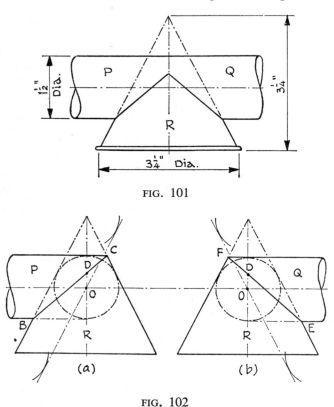

FIG. 101

FIG. 102

full joint line *BC* which would occur if only one part of the cylinder *P* were intended to join the cone *R*. Similarly, the part *DE* is a portion of the full joint line *EF* which would occur if only the part *Q* of the cylinder were intended to join the cone *R*. This principle is illustrated in (*a*) and (*b*), Fig. 102, from which it should be evident that for the whole cylinder the joint line is composed of the two parts *BD* and *DE*.

The pattern for the cone is shown developed in Fig. 103, and is a straightforward example of right conic development. The usual semicircle is described on the base of the cone, and one quadrant is divided into three equal parts. The points are numbered as from 1 to 4, and the middle two are projected perpendicularly back to

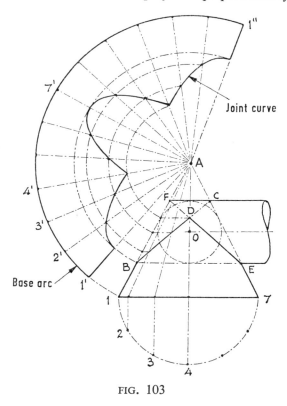

FIG. 103

the base line. From the points on the base line, lines are drawn to the apex *A*, and where they cross the joint line *BD*, the points, including *D*, are projected horizontally to the side of the cone as shown in the Figure. Using the apex *A* as centre, all the points on the side of the cone are now swung into the pattern. Next, the base arc in the pattern is divided into twelve parts equal to those on the quadrant 1 to 4. From all the points on the base arc, radial lines are drawn to the apex *A*. Where the radial lines cross the remaining arcs, points are afforded through which to draw the joint curve in the pattern as shown in Fig. 103.

A RIGHT CONICAL HOPPER ON INCLINED PIPE

The illustration in Fig. 104 represents a right conical hopper fitted on an inclined pipe or duct, and as the joint is composed of two straight lines, it may be assumed that the problem is one of intersection around a common central sphere. The pattern development is shown in Fig. 105, and in general the example is similar to that presented in Fig. 101. First, the cone is drawn to the dimensions given, and the circle representing the sphere is located and drawn as shown in Fig. 105. The joint line *BDE* is next obtained as part of the two separate joint lines *BC* and *EF*.

The semicircle is drawn on the inverted base, divided into six equal parts and the points numbered as from 1 to 7. The points

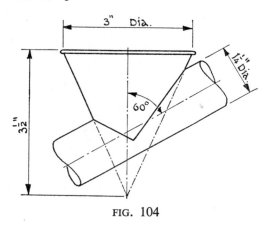

FIG. 104

on the semicircle are dropped perpendicularly on to the base, and from the points on the base, lines are drawn to the apex *A*. Then, from the points where these lines cross the joint line *BDE*, lines are drawn at right angles to the centreline to the side, *A*7, of the cone. Next, using the apex *A* as centre, all the points on the side of the cone are swung into the pattern, and then, beginning at any point 1' on the base arc, twelve divisions are marked off as from 1' to 7' and on to 1", equal to those on the semicircle. From the points on the base arc, radial lines are drawn to the apex *A*.

Before the joint curve in the pattern is drawn, two further points need to be located on the base arc; they are *x'* and *x"*, which, in turn, enable the points *D'* and *D"* to be correctly located. First, from the apex *A*, a line is drawn in the elevation through point *D* to the base of the cone. Then, from this point on the base, and at right angles to the base, a further line is drawn to meet the semicircle

The Common Central Sphere

in point x. Now, the distance from point 3 to x on the semicircle is taken and marked off on the base arc in the pattern from point $3'$ to x'. A similar distance is marked off on the base arc symmetrically from the other end to obtain point x''. Radial lines are now

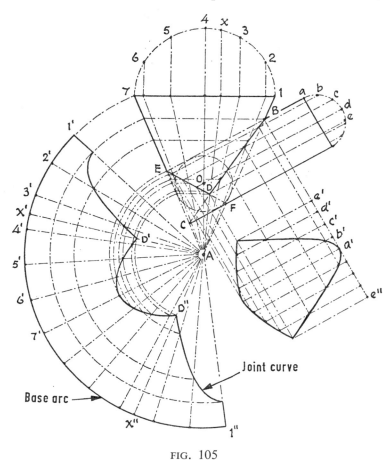

FIG. 105

drawn from these two points to the apex A, and where the radial lines cross the arcs from the corresponding points on the side of the cone, points will be afforded through which to draw the joint curve in the pattern.

The shape of the hole in the cylinder may be obtained by "unrolling" at right angles to the centreline. A semicircle is first drawn on the end of the cylinder, divided into six equal parts, and the points projected back to the end of the cylinder. The usual procedure now is to extend these lines along the cylinder parallel to the centre-line to meet or cut the joint line. In this case, only the lines from the quadrant *abcd* cut the joint line *BDE*. A line from the next point on the semicircle would pass by or miss the point *D*, so a special line is drawn back from point *D* to the semicircle, giving the point *e*. It will be noted that the division *de* on the semicircle is somewhat smaller than the other divisions.

To develop the contour of the hole, lines are projected at right angles to the centreline of the cylinder from the points where the lines on the cylinder cut the joint line *BDE*. Then, along one of these lines, preferably the one projected from point *B*, the divisions *a–b–c–d–e* are marked off as shown in the figure from a' to e' and then repeated in the opposite order from a' to e''. From all the points between e' and e'', lines are now drawn parallel to the cylinder to cut those projected from the joint line. Points are thereby afforded through which to draw the shape of the hole as shown in Fig. 105.

6 Development by Triangulation

The basis of the method of developing patterns by triangulation is to divide the surface of an object into a number of triangles, determine the true size and shape of each, and then lay them down side by side in the correct order to obtain the pattern. The manner in

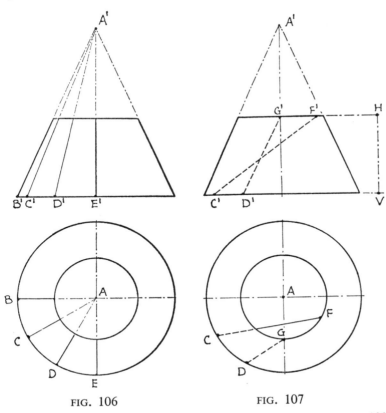

FIG. 106   FIG. 107

which the surface is divided into triangles is of paramount importance, as faulty triangulation produces equally faulty pattern development.

For example, consider the lines on the surface of the right conic frustum in Fig. 106. The lines in the elevation from the points

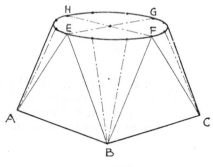

FIG. 108

B', C', D' and E', which all meet at the apex A', lie exactly on the surface of the frustum, as can also be seen in the plan. Now consider the line $C'F'$ in Fig. 107. It will be clear from the plan and elevation that a straight line from C' to F' must pass through space on the inside of the cone and cannot lie on the surface. Similarly,

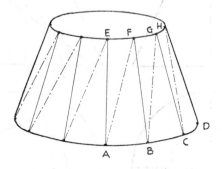

FIG. 109

but to a lesser degree, a straight line from D' to G' will pass up the inside of the cone and not lie on the surface. It should be observed that lines which do lie exactly on the surface are those which pass from points on the base to the apex of the cone. All others which cut somewhat crosswise lie inside the surface to an amount which depends on the degree of deviation from the radial position.

Development by Triangulation

This condition is of considerable importance in dividing the surface of an object into triangles for the purpose of developing the pattern. The aim should be to arrange the triangulation so that as many lines as possible lie exactly on the surface, and those which must lie crosswise should deviate as little as possible from the true surface.

Referring to Fig. 108, which represents a perspective view of an ordinary square-to-circle transforming piece, it will be seen that each corner of the square base is joined to the corresponding quarter of the circle at the top, thereby forming a quarter of an inverted cone between each corner and the top. Also, the straight sides of the square form the bases of flat triangles, as at *ABE* and *BCF*. By these conditions all the lines lie exactly on the surface.

Now, referring to Fig. 109, which represents a perspective view of a right conic frustum, it will be seen that one quarter of the base and the corresponding quarter of the top are divided into the usual three equal parts as shown at *ABCD* and *EFGH*. The series of lines which lie exactly on the surface of the cone are those which occupy the radial positions, such as *AE*, *BF*, *CG* and *DH*. The diagonals *AF*, *BG* and *CH* lie slightly crosswise under the surface and would thereby introduce a slight error in accuracy in triangulating the pattern.

True Lengths by Triangulation

Every line in a plan view is seen as in a horizontal plane. The same line in the elevation, though it may not itself be vertical, may be given a vertical height between its extremities, top and bottom. Referring to Fig. 107, the two lines $C'F'$ and $D'G'$ in the elevation lean or decline at different angles and have different lengths, yet both have the same vertical height, *VH*, between their extremities. Furthermore, neither the length in the plan nor the length in the elevation represents the true length of either of these lines.

The most important rule in triangulation is based on the principle that the plan length of any line when placed at right angles to its vertical height will give a diagonal which is its true length. The true length line is, in effect, the hypotenuse of a right-angled triangle formed by taking the length of a line from the plan and placing it at right angles to its vertical height taken from the elevation. The hypotenuse is then the true length of the line.

TWISTED SQUARE TRANSFORMING PIECE

The square-to-square transforming piece shown in Fig. 110 is perhaps one of the simplest examples by which the principles of triangulation may be illustrated. The surface falls naturally into a series of triangles, and in the construction, two sides of each triangle need to be bent in forming the body.

The development of the pattern is also shown in Fig. 110. The seam is arranged to occur on the flat triangle between points 1 and 2.

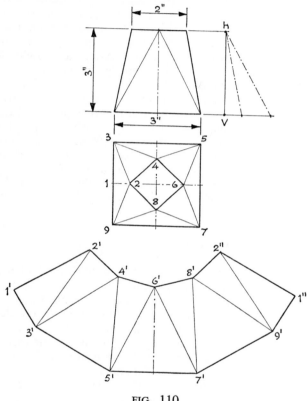

FIG. 110

Then the zigzag line forming the triangles between the top and bottom is numbered 1,2,3,4,5,6,7,8,9,2,1. Since the square at the top is in a horizontal plane parallel to the plane of the square at the bottom, all the diagonals have the same vertical height, VH.

To begin the pattern, the distance from 1 to 2 is taken from the plan and marked off along the base line from V in the elevation. The diagonal line from that point up to point H will give the true length of the line 1,2. This true length is then marked off in the pattern as from 1' to 2'. It may be observed that as the plan line

Development by Triangulation

1,2 is horizontal, its true length is also obtainable from the slant side of the body in the elevation. All the other diagonals, however, must be triangulated against the vertical height in order to obtain their true lengths.

Next, the plan length from 2 to 3 is taken and marked off along the base line from point V. The diagonal up to point H is then taken, and from point 2' in the pattern an arc is described through point 3'. The next line in the pattern, 1'3', is taken direct from the plan. This can be done because the distances round the square, 1,3, 3,5, 5,7 and 9,1, are all in the horizontal plane and have no vertical heights, so that these distances are already true lengths; the same conditions hold good for the distances around the smaller square at the top. Thus, the distance 1 to 3 is taken direct from the plan and from point 1' in the pattern an arc is drawn to mark point 3'.

Now, the plan length 3,4 is taken and triangulated against the vertical height. The diagonal is taken and from point 3' in the pattern an arc is drawn through point 4'. Next, the true distance 2,4 is taken direct from the plan and from point 2' in the pattern an arc is drawn to mark point 4'.

The remainder of the pattern is obtained by repeating this process with the other diagonals and distances round the squares. It will be noted that in this example all the diagonals are equal, having the same plan length and vertical height. It therefore follows that the true lengths are also equal. The exception, in this case, is the joint line from point 1 to point 2, which occurs at each end of the pattern.

SQUARE-TO-CIRCLE TRANSFORMING PIECE

Another example of general importance in triangulation is a transforming piece from a square base to a circular top. Such a transforming piece is illustrated in Fig. 108, and again in Fig. 112. The circular top is usually divided into twelve equal parts, and for purposes of pattern development the straight distances from point to point are taken to space out the length of the top curve in the pattern. It will be clear that this introduces a certain amount of error, though if one disregards practical inaccuracy, the amount of error in the completed pattern is not so great as would at first appear.

For example, consider the part AB of the circle in Fig. 111. That portion of the curve represents one-twelfth of the circumference, or $0.5236r$, where r is the radius of the circle. The straight distance from A to B represents one side of a twelve-sided polygon inscribed in the circle, and is equal to 0·5176 *r*. This amounts to a difference of 0·006 *r* or 1·146 per cent. However, by an inspection of the pattern in Fig. 112 it will be seen that the spaces along the top edge, although marked straight across from point to point, also result in a curve in the pattern. This reduces the small amount of difference which occurs in using the straight line spacings.

To develop the pattern, the plan is divided into triangles as shown in Fig. 112, with the joint or seam arranged to occur on the flat side from 1 to 2. As the top and bottom edges are in horizontal

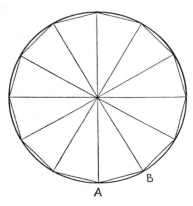

FIG. 111

planes only one vertical height will be needed in developing the pattern. For the same reason also, all the divisions around the circular top and the square bottom present true lengths in the plan, which may therefore be taken direct for use in the pattern.

To obtain the first triangle, the plan length from 1 to 2 is taken and marked off along the base line at right angles to the vertical height *VH*. The true length diagonal is taken and marked off in any convenient position to begin the pattern, as from 1′ to 2′. The next plan length from 2 to 3 is taken and marked off at right angles to *VH*. The true length diagonal is taken, and from point 2′ in the pattern an arc is described through point 3′. Next, the true distance 1,3 is taken direct from the plan, and from point 1′ in the pattern an arc is described to cut the previous arc in point 3′. This completes the first triangle 1′2′3′.

For the second triangle the plan length 3,4 is taken and marked off at right angles to *VH*. The true length diagonal is then taken, and from point 3′ in the pattern an arc is drawn through point 4′. Next, the true distance 2,4 is taken direct from the plan and from

Development by Triangulation

point 2' in the pattern an arc is drawn to cut the previous arc in point 4'. This completes the second triangle.

The third triangle is obtained in a similar manner by taking the plan length 3,5, triangulating it against the vertical height, taking the true length line and marking it off as from 3' to 5' in the pattern.

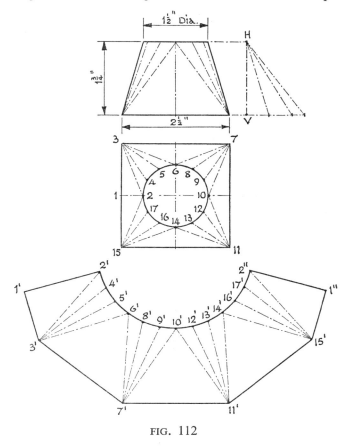

FIG. 112

The true distance 4,5 is next taken from the plan and marked off from point 4' in the pattern, thus completing the third triangle.

The process of obtaining the fourth triangle is again similar to the routine applied to the second and third. The fifth triangle encloses the flat side between points 3, 6 and 7, and the next step is to take the plan length 6,7 and triangulate it against the vertical height. The true length diagonal is then taken and from point 6' in the pattern an arc is described through point 7'. Now, the true distance from 3 to 7 is taken direct from the plan and from point 3' in the pattern an arc is drawn cutting the previous arc in point 7'. That completes the fifth triangle, which is the first full flat side in the development of the pattern.

The remainder of the development is largely a repetition of the development of these five triangles and should readily be followed from the illustration.

SQUARE-TO-CIRCLE TRANSFORMER WITH VERTICAL SIDES

A condition which often presents some difficulty to a beginner is that the transforming piece is to have vertical sides, as in the example shown in Fig. 113. In this case the diameter of the circle is equal to the sides of the square, which brings the four flat triangles on the

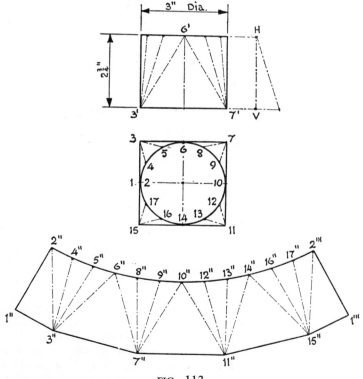

FIG. 113

Development by Triangulation

sides into the vertical position. Thus, the triangle 3'6'7' in the elevation becomes the straight line 3,6,7 in the plan. This condition holds good for the flat triangles on the other three sides as well.

For the development of the pattern the plan is divided into triangles and the points numbered as in Fig. 113. The joint is arranged to occur in the middle of one side from 1 to 2, but as the side is vertical, point 2 lies vertically above point 1, so that in the plan both are represented by the single point 1,2.

To begin the pattern, the first line from 1 to 2 is taken direct from the elevation as the vertical height is its true length. This distance is marked off in the pattern as from 1" to 2". Next, the plan length from 2 to 3 is taken and marked off at right angles to VH, from which the true length diagonal is taken, and from point 2" in the pattern an arc is drawn through point 3". Now, the plan distance from 1 to 3, since it is a part of the square base and is therefore a true length, is next taken, and from point 1" in the pattern an arc is drawn through point 3", thus completing the first triangle.

The next three triangles, 3,2,4, 3,4,5 and 3,5,6, are developed as in the previous example shown in Fig. 112. The fifth triangle is in a vertical plane, and in the plan is seen as the straight line 3,6,7, with points 3 and 7 as the extremities of the base, and point 6 as the apex of the triangle. This triangle is seen in the elevation between the points 3', 6' and 7', which present the full and true size of the triangle. Therefore to add the fifth triangle to the pattern, the side 6' and 7' is taken direct from the elevation, and from point 6" in the pattern an arc is drawn through point 7". Next, the base line 3' to 7' is taken from the elevation, and from point 3" in the pattern an arc is drawn cutting the previous arc in point 7".

The development of the remainder of the pattern should offer no difficulty since it is simply a repetition of what has gone before.

CIRCLE-TO-RECTANGLE NOZZLE

The transforming piece shown in Fig. 114 is a typical example of a nozzle in which a circular top, usually connected to a cylindrical or tubular supply pipe, transforms to a narrow rectangular outlet.

The process of developing the pattern is similar to that applied to the solution of the two previous problems. The seam is arranged to occur on the centreline of one of the end triangles, as from 1 to 2 in the plan. Again, as the top and bottom edges in the elevation are horizontal, only one vertical height will be needed.

To begin the pattern, the plan line from 1 to 2 is taken and marked off at right angles to VH. The true length diagonal is then taken and marked off in any convenient position as from 1' to 2'. It may be noted that this true length is the same as the corresponding side in the elevation, but that is only because the line 1 to 2 is horizontal in the plan. All others must be triangulated against the vertical height. Next, the plan length 2 to 3 is taken and marked off at right angles to *VH*. The true length diagonal is taken and from point 2' in the pattern an arc is drawn through point 3'. Now, since the rectangular bottom is horizontal, the next length from point 1

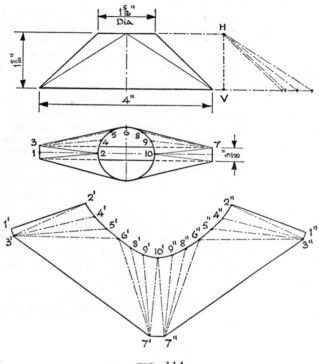

FIG. 114

to 3 in the plan is taken as a true length and from point 1' in the pattern an arc is drawn through point 3', thus completing the first triangle.

The development of the next three triangles, 3,2,4, 3,4,5 and 3,5,6, follows the same routine as that applied to the corresponding triangles in Figs. 112 and 113. The fifth triangle is obtained by taking the plan length 6 to 7, triangulating it against the vertical height from which the true length diagonal is then taken, and drawing an arc through point 7' from point 6' in the pattern. Then the

Development by Triangulation

true plan distance from 3 to 7 is taken and from point 3' in the pattern an arc is drawn through point 7'.

The remainder of the pattern should now be readily followed from this point, as the directions are merely a repetition of terms.

A TRANSFORMER FROM RECTANGLE–SEMI-ELLIPSE TO CIRCLE

The example shown in Fig. 115 represents a hood with a circular top which may be connected to a cylindrical uptake pipe. The back

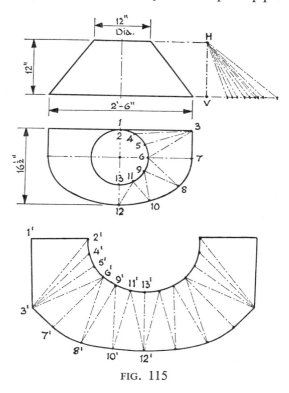

FIG. 115

of the hood is to fit against a wall, and the body is constructed of half a rectangle at the bottom which transforms to the corresponding half of the circular top. The front embodies a transforming piece between half an ellipse at the base to the corresponding half of the circle at the top.

The seam is arranged to occur at the middle of the vertical back, as from point 1 to 2. It should be observed that as the seam, 1 to 2, is vertical, point 2 in the top is directly above point 1 in the bottom, and both points are therefore represented by the single point 1,2 in the plan. In dividing the surface into triangles, the first part from point 1 to point 7 is treated as a rectangle-to-circle transformer. The second part from point 7 to point 12 is triangulated between

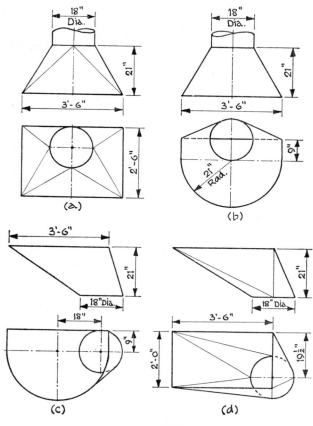

FIG. 116

the quarter ellipse at the bottom and the corresponding quarter circle at the top. In triangulating the elliptical part it should be noted that the points on the ellipse are first joined to the corresponding points on the circle, as shown in the right-hand side of the plan, Fig. 115, and then the diagonals are placed across the shorter distance between the opposite points in each section. This ensures

Development by Triangulation

the minimum of error in the development of the pattern, as explained in connexion with Fig. 107.

To begin the pattern, the first line from 1 to 2 is taken direct from the elevation as that is its true length, and then marked off in any convenient position in the pattern, as shown from 1' to 2'. Next, the plan length from 2 to 3 is taken and marked off at right angles to the vertical height VH. The true length diagonal is then taken and from point 2' in the pattern an arc is drawn through point 3'. Now, the true horizontal distance from point 1 to point 3 is taken direct from the plan and from point 1' in the pattern an arc is drawn through point 3', thus completing the first triangle. The next four triangles, 3'2'4', 3'4'5', 3'5'6' and 3'6'7', should be readily followed from the illustration on the basis of directions given in previous examples. The next triangle 7'6'8' is obtained by triangulating the plan length 6,8 against the vertical height. The true length diagonal is then taken and from point 6' in the pattern an arc is drawn through point 8'. Then the distance between points 7 and 8 is taken direct from the plan, and from point 7' in the pattern an arc is drawn through the previous arc to locate point 8'.

This process should now be repeated with all the diagonals and distances between point 8 and point 13, which then completes half the pattern to point 13'. The remaining half is, of course, a repetition of the first half but in the reverse order.

Further Examples for Practice The examples of transforming pieces given in Fig. 116 for further practice are typical of the wide variety which are used in industry for innumerable purposes. Some of the chief uses are for hoppers, hoods, chutes and ductwork. Transforming pieces may have any shape at either end. The examples given in Fig. 116 have tops and bottoms parallel and in horizontal planes, which means that only one vertical height will be needed for the development of the patterns. Figs. 116(*a*) and (*b*) represent typical hoods and Figs. 116(*c*) and (*d*) are representative of hopper construction.

SQUARE-TO-CIRCLE TRANSFORMER WITH INCLINED CIRCULAR TOP

The examples of development by triangulation so far considered have been such that the top and bottom edges lay along two parallel planes. This condition involves the use of only one vertical height. Most transforming pieces of complex design do t nofollow so simple a pattern, but are such that many vertical heights are involved, often one for every line. The example given in Fig. 117 has a circular top which inclines at an angle to the horizontal. The solution of this problem requires a series of vertical heights corresponding to the points on the inclined top.

Preparatory to dividing the surface into triangles the circular top must first be dropped into the plan as an ellipse. A semicircle is

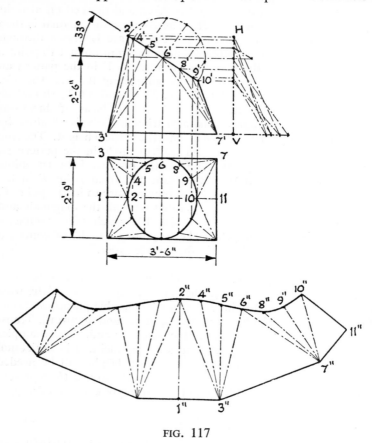

FIG. 117

described on the diameter in the elevation, divided into the usual six equal parts and the points projected back at right angles to the top edge. From the points on the top edge, lines are drawn vertically downwards into the plan. Then distances equal to the corresponding widths of the semicircle on the top edge, are marked off above and below the horizontal centreline in the plan. Thus points are obtained on the ellipse to enable the curve to be drawn in. Since the points

Development by Triangulation

on the ellipse correspond to those on the top edge in the elevation, they may now be used in dividing the surface of the transforming piece into triangles as shown in Fig. 117.

Prior to developing the pattern, a vertical height line VH is first erected, and all the points on the inclined top edge are projected horizontally to VH to establish their vertical heights thereon. The seam is arranged to occur on the short side between points 10 and 11 as seen in the plan. Since the body is symmetrical about the horizontal centreline in the plan, the pattern is begun from the opposite side, as from 1 to 2, and both sides are developed symmetrically so that the ends of the pattern, 10 to 11, form the joint when the body is shaped to the required form.

To begin the pattern, the plan length 1 to 2 is taken and marked off along the base line at right angles to VH. The true length diagonal is taken up to the top point on VH level with point 2' in the elevation and marked off in any convenient position in the pattern, as from 1″ to 2″. Next, the plan length 2 to 3 is taken and marked off at right angles to VH. The true length diagonal is taken again up to the top point level with 2' in the elevation, and from point 2″ in the pattern an arc is drawn through point 3″. Now, the true distance from point 1 to point 3 is taken direct from the plan and from point 1″ in the pattern an arc is drawn cutting the previous arc in point 3″, thus completing the first triangle.

For the second triangle the plan length 3,4 is taken and marked off at right angles to VH. The true length diagonal is taken this time up to the point on VH level with point 4' in the elevation, and from point 3″ in the pattern an arc is drawn through point 4″. The next distance from 2″ to 4″ cannot be taken from the plan as in previous examples, since the spacings around the ellipse are not true distances. The true spacings for the top circular edge are those around the semicircle which is divided equally into six parts. Therefore, one of those divisions is taken and from point 2″ in the pattern an arc is drawn cutting the previous arc in point 4″.

The next two triangles 3″4″5″ and 3″5″6″ are developed in a similar manner, taking care that the correct vertical heights are used up to the points level with 5' and 6' in the elevation, and that the spacings for the top edge are taken from the semicircle and not from the plan.

The fifth triangle is in a vertical plane. In the plan it is represented by the straight line 3,6,7, and in the elevation by the triangle 3'6'7', which, in the latter view, represents its true size and shape. Therefore the true lengths required for the pattern are taken direct from the elevation. Thus, the line 6' to 7' is taken, and from point 6″ in the pattern an arc is drawn through point 7″. Next, the true distance 3'7' from the elevation or the distance 3,7 from the plan is taken and from point 3″ in the pattern an arc is drawn to cut the previous arc in point 7″. This completes the fifth triangle.

The remaining triangles, 7,6,8, 7,8,9, 7,9,10 and 7,10,11, may readily be added from this point. The chief precautions are to make sure that the correct vertical heights are used, and that the spacings for the top edge are taken from the semicircle in the elevation. The directions given above apply to one half of the pattern as from point 1 to point 11 in the plan. The full pattern is most conveniently

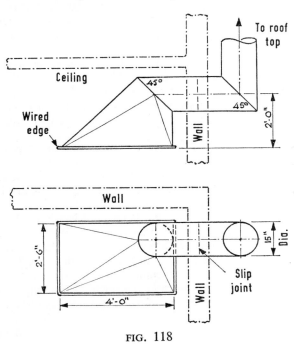

FIG. 118

developed by repeating or duplicating the process on the other side of the centreline 1″2″ as each line or spacing is obtained.

An alternative method of obtaining the true spacings from 2″ to 10″ in the pattern, is to triangulate each plan spacing against its appropriate vertical height. For example, the plan distance 5,6 may be taken and marked off from the vertical height line opposite to the point 6′ in the elevation. The diagonal taken up to the point level with 5′ will then give the true length spacing between those two points. However, in the present case it is much more convenient to take the true spacings from the semicircle drawn on the top edge.

Development by Triangulation 125

FUME HOOD AND PIPE

Fig. 118 represents a practical application of an off-centre transforming piece with an inclined top edge. It consists of a hood to take away fumes from a vat placed in the corner of a room. The hood is connected to a pipe which passes horizontally through a wall and turns upward to carry the fumes to roof level. The development of the hood is shown in Fig. 120, and the development of the off-set pipe connexion is similar to that of the example shown in Fig. 74.

An important condition arising from the arrangement of this piece of work is the shape of the joint line between the top of the hood and the horizontal pipe through the wall. Since the pipe is of right

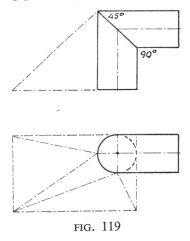

FIG. 119

cylindrical construction and the joint line is at 45° to its axis, it follows that the true shape of the joint line is an ellipse. Also, since the central axis of the pipe is horizontal and the joint line is at 45°, it follows that the joint line is seen as a circle on the plan. This condition is shown in Fig. 119, which represents a right-angled elbow with one limb horizontal and the other vertical. The plan view of the vertical limb would present a circle, and if this limb were removed, leaving the horizontal pipe in position, the joint line would still present a circle in the plan. The addition of a hood in place of the vertical limb does not alter this condition.

Prior to developing the pattern, Fig. 120, the plan is divided into triangles in a similar manner to that in Fig. 117. The chief difference is that this transforming piece is not symmetrical about any centre-line and therefore the whole pattern has to be triangulated for development. The seam is arranged to occur on the vertical side from point 1 to point 2. Again it will be observed that point 2 lies vertically above point 1 and that both are represented by a single point in the plan.

Another preliminary is to plot the semi-ellipse on the joint line in the elevation. To do this, the circle in the plan is divided into twelve equal parts and the points thereon are projected vertically upwards to the joint line in the elevation. From the points on the joint line, lines are drawn at right angles to it, and marked off equal in length to the corresponding semi-width of the circle in the plan. Thus, points are obtained through which to draw the semi-ellipse in the elevation, which represents the true shape of half of the top edge of the transforming piece. The spaces around the semi-ellipse are therefore the correct true distances required for the pattern.

To develop the pattern, the first line 1' to 2' is taken direct from the elevation since its vertical height is its true length, and marked off in any convenient position, as from 1" to 2". Next, the plan length from 2 to 3 is taken and marked off at right angles to VH. The true length diagonal is taken and from point 2" in the pattern an arc is drawn through point 3". Then, the true plan length 1,3 is taken and from point 1" in the pattern an arc is drawn to cut the previous arc in point 3".

For the second triangle, the plan length 3,4 is taken and marked off at right angles to VH. The true length diagonal is taken up to the point on VH which is level with point 4' on the joint line, and from point 3" in the pattern an arc is drawn through point 4". The next spacing in the pattern is *not* obtained from the circle in the plan, but is taken from the corresponding spacing on the ellipse in the elevation. It should be noted that the spacings on the ellipse are not equal, and it is therefore important that the correct order of spacings should be taken for use in developing the pattern. Thus, the spacing on the ellipse which corresponds to that in the plan from point 2 to point 4 is taken, and from point 2" in the pattern an arc is drawn cutting the previous arc in point 4".

For the third triangle, the plan length 3,5 is taken and marked off at right angles to VH. The true length diagonal is taken up to the point on VH which is level with point 5' on the joint line, and from point 3" in the pattern an arc is drawn through point 5". The next spacing, from 4" to 5" in the pattern, is taken from the ellipse in the elevation, and is the distance between the points on the ellipse, which correspond to 4 and 5. The distance is then marked off from point 4" in the pattern by describing an arc which cuts the previous arc in point 5".

For the fourth triangle the plan length 3,6 is taken and marked off at right angles to VH. The true length is taken up to the point

Development by Triangulation

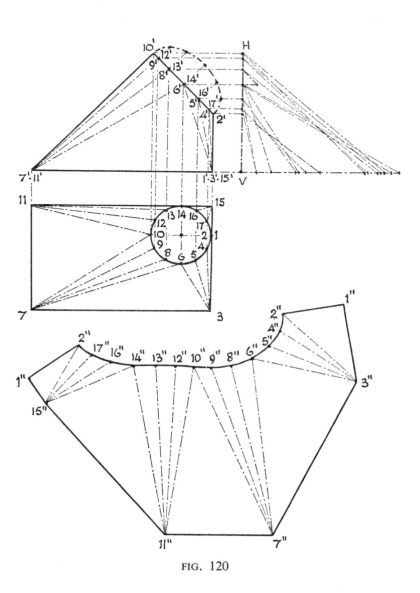

FIG. 120

on VH which is level with point 6′ on the joint line, and from point 3″ in the pattern an arc is drawn through point 6″. The next spacing is again taken from the ellipse between the points corresponding to 5 and 6, and from point 5″ in the pattern an arc is drawn cutting the previous arc in point 6″.

For the fifth triangle the plan length 6,7 is taken and marked off at right angles to VH. The true length diagonal is taken up to the point on VH level with point 6′ on the joint line, and from point 6″ in the pattern an arc is drawn through point 7″. Now, the next distance 3 to 7 is horizontal and is therefore already a true length. Thus, the length 3 to 7 is taken direct from the plan and from point 3″ in the pattern an arc is drawn cutting the previous arc in point 7″.

The remainder of the pattern may be developed from this point by taking each line between the top and bottom edges and triangulating it against the appropriate vertical height. The precautions necessary for the successful completion of the pattern are to see that the correct vertical heights are used, that the correct spacings are taken from the ellipse, and that the distances around the rectangular base are taken as true lengths.

The development of the hood shown in Fig. 120 is a further example wherein the true spacings along the top edge of the pattern may be obtained by triangulating the plan spacings against their corresponding vertical heights. For example, the plan length 13,14 may be taken and marked off from the vertical height line along the line level with point 14′ in the elevation. The diagonal taken up to the point level with 13′ will then give the true distance between points 13 and 14. This distance should be equal to the corresponding spacing on the semi-ellipse at the top of the hood.

Although the true distances of all the spacings around the elliptical top may be obtained in the same way, the method of plotting the semi-ellipse is to be preferred as it becomes evident at a glance how and where the true spacings are obtained for the pattern.

TRANSFORMER FROM RECTANGULAR SLOT TO CIRCLE

Sometimes a transforming piece takes the shape of a nozzle connexion. An example of this kind is shown in Fig. 121, where a cylindrical air duct is connected by a transformer to a rectangular slot in the side of a machine framework. If this illustration were turned sideways, that is, through 90°, so that the side elevation became the plan, then the process of developing the pattern would be similar to that of the transforming piece shown in Fig. 120. However, sometimes the process may be somewhat simplified by using an auxiliary projection as the plan.

Development by Triangulation

The development of the pattern by using an auxiliary projection is shown in Fig. 122. The advantage of this method lies in reducing the number of vertical heights needed for the development. In this case only two vertical heights are required instead of seven, and the projected elliptical joint line gives true spacings for the pattern direct from the new "plan." Regarding the projection as the plan view for purposes of development, then the "vertical" heights will be taken in the direction of the projection, as indicated by the line *VH* in Fig. 122.

To plot the ellipse in the auxiliary projection, a semicircle is drawn on the top of the cylindrical uptake pipe in the elevation,

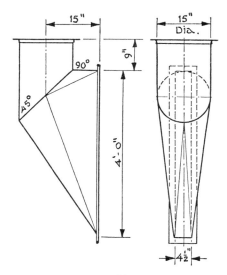

FIG. 121

and divided into six equal parts. The points thereon are projected perpendicularly back to the top of the pipe and onward to the joint line *AB*. From the points thus obtained on the joint line, lines are drawn at right angles to it into the projection. A centreline is then drawn in any convenient position at right angles to the projection lines, as at *CL*. Next, the various widths of the semicircle are taken from the top edge of the pipe, and marked off on the corresponding projection lines on both sides of the centreline *CL*. The ellipse, then, is the true shape of the joint line, and the spacings for the pattern may be taken from its perimeter. To complete the projected view,

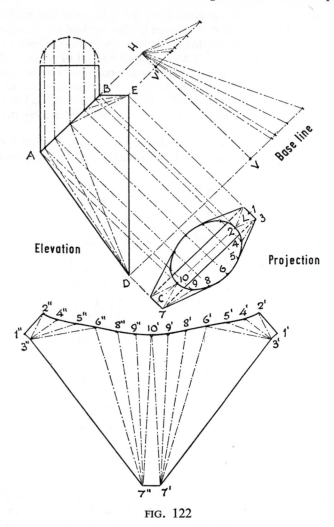

FIG. 122

lines are drawn into the projection from the rectangular edge *DE*, and its width, taken from the side elevation, is marked centrally across the centreline *CL*. The two sides of the rectangle are then drawn parallel to the centreline. From the corners of the rectangle, lines are drawn as tangents to the ellipse, which then completes the projection.

Development by Triangulation

For the development of the pattern the projection is now used as the plan, and is divided into triangles as shown in Fig. 122. The vertical height line, VH, is erected from the base line drawn from point D at right angles to the projection lines. Point E is also projected parallel to the base line to cut VH in V' and provide a second vertical height from V' to H. The seam is arranged to occur down the middle of the small triangle at BE.

The first line in the pattern, from point 1 to 2, may be triangulated against its vertical height or taken direct from the elevation since, in this case, the line BE is its true length. This distance is then marked off in any convenient position to begin the pattern, as from $1'$ to $2'$. Next, the plan length from 2 to 3 is taken and marked off at right angles to the vertical height $V'H$. The true length diagonal up to H is taken and from point $2'$ in the pattern an arc is drawn through point $3'$. The true distance from 1 to 3 is now taken direct from the plan, and from point $1'$ in the pattern an arc is drawn cutting the previous arc in point $3'$.

For the second triangle, the plan length 3,4 is taken and marked off at right angles to $V'H$. The true length diagonal is then taken, and from point $3'$ in the pattern an arc is drawn through point $4'$. The next distance, from point 2 to 4, is taken direct from the ellipse in the plan, and from point $2'$ in the pattern an arc is drawn cutting the previous arc in point $4'$. The development of the next two triangles, $3'4'5'$ and $3'5'6'$, is similar to that of the second triangle $3'2'4'$.

For the fifth triangle, the plan length from 6 to 7 is taken and marked off, this time along the bottom base line from V at right angles to VH. The true length diagonal up to the top point H is taken, and from point $6'$ in the pattern an arc is drawn through point $7'$. Next, the true distance from D to E is taken direct from the elevation, and from point $3'$ in the pattern an arc is drawn cutting the previous arc in point $7'$. It should be specially noted that the edge, DE, of the rectangle in the elevation is the true length of the line 3,7. It is a common error to take the plan lengths 3,7 and use it as the true length in the pattern.

From this point the remainder of the pattern should offer no difficulty since the process of development is similar to that described above. Two precautions must be observed, however. The first is to use the full vertical height VH when triangulating the lines radiating from the bottom corners of the rectangle at D in the elevation, and to use the smaller vertical height $V'H$ when triangulating the lines radiating from the top corners of the rectangle at E in the elevation. The second precaution is to use the appropriate divisions from the ellipse in the plan for the true spacings of that edge in the pattern.

RECTANGLE-TO-CIRCLE HOPPER WITH INCLINED BACK

The example shown in Fig. 123 represents a hopper with a circular hole at the bottom, and rectangular top with the rear half of the rectangle inclined upwards to form a raised back. The development of the pattern is shown in Fig. 124, in which one half of the plan is divided up for triangulation. The numbering of the points begins at the back of the hopper as from 1 to 2, so that the two halves of

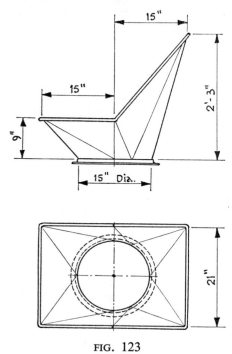

FIG. 123

the pattern may be developed symmetrically on both sides of the line 1″ to 2″. It will be seen that the two vertical heights will be needed as shown at VH and $V'H'$. The second height $V'H'$ is shown on the left of the elevation, though it might alternatively be marked off and used on the first height VH.

To begin the pattern, the plan distance between points 1 and 2 is taken and marked off along the base line level with point V. The true length diagonal is taken up to the point H and the first line 1″ to 2″ is marked off in any convenient position. It may be noted that as the plan line 1 to 2 is horizontal, its true length could be

Development by Triangulation

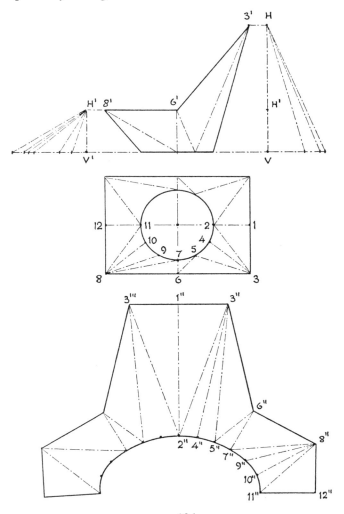

FIG. 124

taken direct from the corresponding points in the elevation. Next, the true length line from 1 to 3 is taken direct from the plan, and from point 1″ in the pattern an arc is drawn through points 3″ and 3‴ on either side of point 1″, and the line 3″3‴ is drawn at right angles to the first line 1″2″. Lines are now drawn from points 3″ and 3‴ to point 2″, which then completes the first triangle.

To proceed with the right-hand side of the pattern, the plan length from 3 to 4 is taken and marked off at right angles to the vertical height VH. The true length diagonal is taken, and from point $3''$ in the pattern an arc is drawn through point $4''$. Next, the spacing 2 to 4 is taken direct from the plan, and from point $2''$ in the pattern an arc is drawn cutting the previous arc in point $4''$, thus completing the second triangle.

For the third triangle, the plan length 3,5 is taken and marked off at right angles to the vertical height VH. The true length diagonal is taken and from point $3''$ in the pattern an arc is drawn through point $5''$. Next, the true spacing 4 to 5 is taken direct from the plan, and from point $4''$ in the pattern an arc is drawn cutting the previous arc in point $5''$.

For the fourth triangle, the plan length 5,6 is taken and marked off at right angles to the lower vertical height $V'H'$. The true length diagonal is taken, and from point $5''$ in the pattern an arc is drawn through point $6''$. Next, since the plan line 3,6 is horizontal, its true length may be taken direct from the elevation, as from $3'$ to $6'$, and from point $3''$ in the pattern an arc is drawn through point $6''$, thus completing the fourth triangle.

The development of the remaining triangles should be readily followed from this point as the process is the same round to the points $11''$ and $12''$. Since the pattern is symmetrical about the centreline $1''2''$, the opposite half on the left-hand side may be developed simultaneously with the right-hand side.

TRANSFORMER CONNEXION TO ANGULAR CORNER

The transformer connexion shown in Fig. 125 is designed to fit on an angular corner, or to form a ventilator outlet on a roof top. There is one point of geometrical interest connected with this type of problem which might be considered before turning to the development of the pattern.

Referring to Fig. 126, it will be seen that the triangle 3,5,6 in the plan has its apex at point 5 and not at the usual point 7 which lies on the extremity of the quadrant 2,7. The same triangle in the elevation occurs at $3'5'6'$, with its apex at $5'$.

Consider, now, a projection of this triangle and the circular top as seen in the direction of its base line $6'3'$. This projection is represented diagonally below the elevation, and the triangle becomes a single line in which the points $3'$ and $6'$ occupy the single point shown at $3^0,6^0$, while point $5'$ occurs at the point 5^0 in the projection.

Now imagine the triangle $3'5'6'$ to be a flat surface which may be revolved or rocked on its base line $3'6'$. In the projected view

Development by Triangulation

this triangle, represented by the straight line $3^05^06^0$, would revolve around the single point $3^0,6^0$, and come to rest against the ellipse at point 5^0, which is the nearest point at which the line forms a tangent to the ellipse. Thus it will be seen that point 5 is the most suitable position for the apex of the triangle 3,5,6.

Half of the plan is divided up for triangulation with the points numbered so that the seam falls at the middle of the top triangle

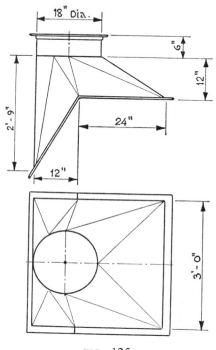

FIG. 125

from 11 to 12. Thus, the first line from 1″ to 2″ will form the centreline of the pattern since the body is symmetrical about the horizontal centreline 1,12. Two vertical heights will be needed, as shown at VH and $V'H'$. The second height, $V'H'$, as in the previous example, might well be marked off and used on the first height VH.

The first line in the pattern may be obtained direct from the elevation, from 1′ to 2′, since that is already its true length, and marked off in any convenient position as from 1″ to 2″. Next, the plan distance from 1 to 3 is taken and marked off on both sides of point 1″ in the pattern. The line 3″ to 3‴ is then drawn at right angles

136 Sheet Metal Drawing and Pattern Development

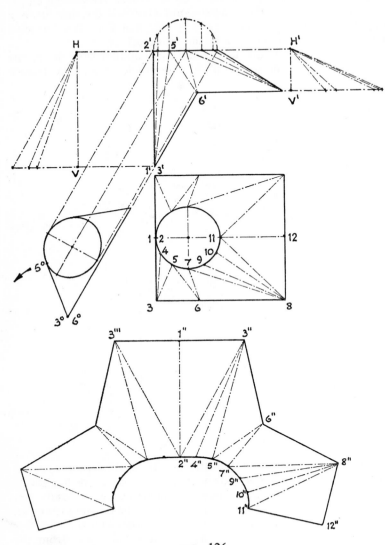

FIG. 126

Development by Triangulation 137

to 1″2″, and the ends joined to point 2″, thus completing the first triangle.

For the second triangle the plan length 3,4 is taken and marked off along the base line from V at right angles to VH. The true length diagonal is taken and from point 3″ in the pattern an arc is drawn through point 4″. Next, the true spacing from 2 to 4 is taken direct from the plan and from point 2″ in the pattern an arc is drawn cutting the previous arc in point 4″.

For the third triangle the plan length 3,5 is taken and marked off along the base line from V. The true length diagonal is taken and from point 3″ in the pattern an arc is drawn through point 5″. Next, the true spacing from 4 to 5 is taken direct from the plan and from point 4″ in the pattern an arc is drawn cutting the previous arc in point 5″.

For the fourth triangle the plan length 5,6 is taken and marked off along the base line, this time from point V', the shorter vertical height. The true length diagonal is taken and from point 5″ in the pattern an arc is drawn through point 6″. Next, the true distance from 3′ to 6′ is taken, this time from the elevation, and from point 3″ in the pattern an arc is drawn cutting the previous arc in point 6″. This completes the fourth triangle.

For the remaining triangles the same routine is followed. The only precaution to observe from this point is that all the lines between the top and bottom have the shorter vertical height $V'H'$. Also, since the pattern is symmetrical about the centreline 1″2″, the opposite half on the left-hand side will be a duplicate of that on the right-hand side.

STOVE ELBOW CONNEXIONS

The shape of fume outlets from stoves of various kinds often takes the form of a rectangular stump with semicircular ends. The outlet may discharge directly into a chimney, or a sheet-iron connecting piece may be made from the outlet to a chimney or to a cylindrical flue which may conduct the fumes to outside atmosphere. The connecting piece from the stove outlet usually consists of a short sleeve to fit closely round the outlet and a transforming piece which connects the sleeve to a cylindrical pipe. This in turn either serves directly as a flue or chimney, or connects the transforming piece to a chimney. Thus the connecting units give rise to a variety of problems of pattern development according to the conditions existing in any particular case.

The examples shown in Fig. 127 represent four different sets of conditions. In (*a*) and (*b*) two outlets are connected horizontally while (*c*) and (*d*) present vertical connexions. The cylindrical pipes are either on centre or off centre and may off-take at any angle from the transforming piece. These are only a few of the many examples which could be given.

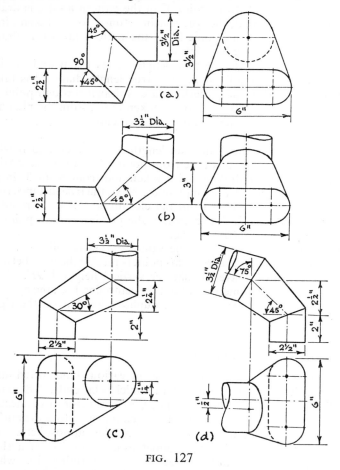

FIG. 127

The patterns for the example in Fig. 127(*c*) are shown developed in Fig. 128. An auxiliary projection is used as a "new plan" in order to simplify the question of vertical heights. It will be seen that as the two joint lines are in parallel planes, and the projection is made at right angles to those planes, only one "vertical" height is needed for the development of the pattern. If the ordinary plan

Development by Triangulation

were used, then each and every line in the triangulation would require its own vertical height, and all of them would be different. Therefore, by using the auxiliary projection as a "new plan" on section *A*, the line *VH* at right angles to the new base line may be used as the "vertical" height between the two joint lines.

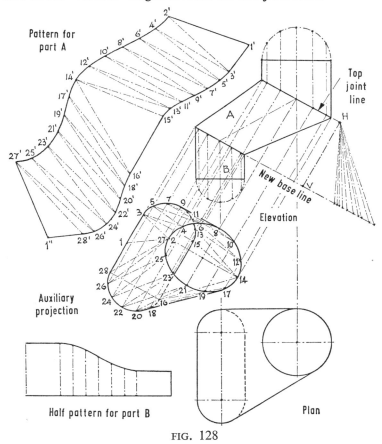

FIG. 128

It is assumed that by this stage the reader is familiar with the method of making the auxiliary projection and will be able to follow it from the illustration, as there are no special principles involved such as will be discussed in a later chapter.

The top joint line, which cuts the cylindrical up-take pipe at an angle, presents an ellipse in the projection. The bottom joint line, which cuts the vertical outlet stump at an angle, presents a shape similar to that shown in the plan, except that its width is increased to correspond to that of the joint line in the elevation, and its ends are semi-ellipses instead of semicircles. It should be noted that the divisions obtained around the ellipses in the auxiliary projection are true spacings. These true spaces are necessary for the development of the pattern.

Referring to Fig. 128, the joint line in the transforming piece is arranged to occur as from point 1 to 2, where point 1 falls at the middle of the straight line 3,28 and point 2 at the corresponding end of the ellipse. The surface is then triangulated by numbering the points in zigzag fashion from top to bottom, as shown by the numbers 1,2,3,4 . . . 27,28, and back to number 1.

To begin the development of the pattern, the plan length from point 1 to point 2 is taken and marked off from point V along the new base line. The true length diagonal up to point H is then taken and marked off in any convenient position, as from 1' to 2' in the pattern. Next, the plan length from 2 to 3 is taken and marked off from V along the new base line. The true length diagonal up to point H is taken and from point 2' in the pattern an arc is drawn through point 3'. Next, the plan length from 1 to 3, which is already a true length, is taken and from point 1' in the pattern an arc is drawn cutting the previous arc in point 3'. Lines joining these points complete the first triangle.

For the second triangle the plan length from 3 to 4 is taken and marked off from V along the new base line. The true length diagonal up to point H is taken, and from point 3' in the pattern an arc is drawn through point 4'. The next distance from 2 to 4 is taken direct from the ellipse in the plan, and from point 2' in the pattern an arc is described cutting the previous arc in point 4'. This completes the points required for the second triangle.

For the third triangle the plan length is taken from 4 to 5 and marked off at right angles to the vertical height. The true length diagonal is taken, and from point 4' in the pattern an arc is drawn through point 5'. Next, the distance 3 to 5 is taken direct from the plan, and from point 3' in the pattern an arc is drawn cutting the previous arc in point 5'.

From this point the development of the remainder of the pattern should be straightforward, since the directions would be merely a repetition of terms using the appropriate point numbers. The chief precautions to observe are that the spacings around the ellipse and the distances around the bottom shape in the plan are already true distances, whilst all the plan lengths taken between the top and bottom edges must be triangulated against the vertical height in order to obtain their true lengths.

Development by Triangulation

A half pattern for the vertical stump B is also shown in Fig. 128, obtained by simple parallel line development.

Breeches Pieces

The term "breeches piece" is generally applied to that type of branch piece in which the joint between the branches is curved, and differs in that respect from the "juntion piece" which has flat triangular pieces between the branches. The junction piece is dealt with on p. 159.

In the case of the breeches piece the form of the joint between the branches may be made to any desired shape, though in cases where both branches are drawn up from the same base, as in Fig. 129, the joint generally conforms to a definite shape dependent on the diameters of the branches and the angle between them.

BREECHES WITH BRANCHES ON SAME BASE

The example shown in Fig. 129 represents a breeches piece in which both branches, if produced to the base line, would terminate

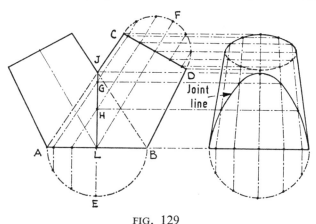

FIG. 129

on the same circular base AB. In this case the joint line JL conforms to the definite shape shown in the side elevation, which is of the right-hand limb only, thus showing the joint line in full.

To obtain the shape of the joint line, the semicircles drawn on the diameters AB and CD are each divided into six equal parts, and the points thereon projected back and at right angles to the respective diameters. The points on the diameters obtained from the quadrants *AE* and *CF* are now joined by lines which cross the joint line *JL* at points *G* and *H*.

In the side elevation, a semicircle is drawn on the base to correspond to that on *AB*, and is divided into six equal parts. The points on the semicircle are then projected upwards perpendicular to the base. Next, the circular top edge *CD* is projected into the side elevation to obtain the ellipse in that view. The points around the top half of the ellipse are now joined to the corresponding points along the base line in the side elevation, as shown in Fig. 129.

Next, the points *J*, *G* and *H* are projected horizontally from the front elevation to meet the corresponding lines in the side elevation, thereby affording points through which to draw the elliptical shape of the joint line. Thus, by this method of constructing a breeches piece the shape of the joint line is fixed.

The development of the pattern is shown in Fig. 130. A plan of the right-hand limb is drawn, in which the circular top becomes an ellipse in the plan. Only the bottom half of the plan is triangulated for the development. In dividing the surface into triangles, the semicircle in the plan is divided into six equal parts and lines are drawn from the points thereon to the corresponding points on the ellipse, as from 17 to 13, 16 to 11, 15 to 9, 8 to 7, 6 to 5, 4 to 3, and 2 to 1. The first three of these lines cross the joint line in the plan at points 10, 12 and 14. Diagonals are now drawn from 2 to 3, 4 to 5, 6 to 7, 8 to 9, 10 to 11, and 12 to 13, thus dividing the surface of the branch piece into a series of triangles.

Prior to developing the pattern, the true shape of one half of the joint line is plotted in the elevation, as shown by the curve from point 14' to point 17'. This curve is similar to one half of that shown in the side elevation, Fig. 129, and serves the purpose of presenting true spacings for the pattern. To obtain the curve in the elevation, the points 15 and 16 are projected vertically upwards from the plan to points 15' and 16' on the base line in the elevation. Then, lines are drawn from points 15' and 16' to points 9' and 11' on the top edge of the limb. These two lines cross the joint line at points 10' and 12'.

Imagine, now, that the joint line 8 to 14 in the plan, and 8' to 14' in the elevation, is revolved on its vertical axis through 90° to produce the true curve from 14' to 17'. To obtain this curve, the points 10 and 12 in the plan are swung round from the centrepoint 14 to the horizontal centreline 14,17. From the points on 14,17 lines are drawn vertically upwards to meet the horizontal lines drawn from points 10' and 12' in the elevation. Thus, points 10^0 and 12^0 are obtained through which to draw the curve from 14' to 17'.

Development by Triangulation

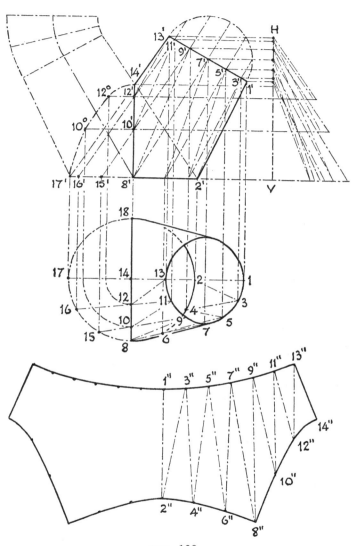

FIG. 130

The development of the pattern is obtained by straightforward triangulation. The points along the top edge 1' to 13' are projected horizontally to the vertical height line VH, as also are the points 10', 12' and 14' to serve as base lines in obtaining true lengths.

The first line for the pattern, 1 to 2, may be obtained direct from the elevation, since the line 1' to 2' is already a true length. Alternatively it may be obtained by taking the plan length 1 to 2 and triangulating it against its vertical height, which in this case gives the same length. Thus, the line 1' to 2' is taken and marked off in any convenient position, as at 1" to 2", to begin the pattern. Next, the plan length 2 to 3 is taken and marked off along the base line from point V, the true length diagonal taken up to the point level with 3', and from point 2" in the pattern an arc is drawn through point 3". To complete the first triangle, the true spacing is needed from 1" to 3". This spacing is taken, not from the ellipse in the plan, but from the semicircle on the top edge 1'13' in the elevation. The divisions around the semicircle are the true spacings required for the pattern. Therefore one of these divisions is taken, and from point 1" in the pattern an arc is drawn cutting the previous arc in point 3".

For the second triangle the plan length 3 to 4 is taken and triangulated against the vertical height. The true length diagonal is taken up to the point level with 3', and from point 3" in the pattern an arc is drawn through point 4". Now the true distance from 2 to 4 is taken direct from the plan, and from point 2" in the pattern an arc is drawn cutting the previous arc in point 4".

For the third triangle the plan length 4 to 5 is taken and marked off along the base line from V. The true length diagonal is taken, this time up to the point level with 5', and from point 4" in the pattern an arc is drawn through point 5". The next true spacing from 3" to 5" is again taken from the semicircle on the top edge, and from point 3" in the pattern an arc is drawn cutting the previous arc in point 5".

The remainder of the pattern should be readily followed from this point, since the process is similar throughout. The chief points to observe are that the spacings along the top edge in the pattern from point 1" to 13" are taken from the semicircle in the elevation; the spacings in the pattern from 2" to 8" are taken direct from the corresponding quadrant 2 to 8 in the plan; and the spacings in the pattern from 8" to 14" are taken from the corresponding distances along the true joint curve 17' to 14' in the elevation. All the zigzag lines in the pattern are obtained by triangulating the plan length of each line against its respective vertical height. For example, consider the line from point 10 to point 11. This line is taken from the plan and marked off at right angles to the vertical height, not

Development by Triangulation

along the bottom base line from V, but along the line which is level with point 10′ in the elevation. The true length diagonal is then taken from that point up to the point level with 11′. Again, the plan length 11 to 12 must be marked off at right angles to the vertical height, but this time along the line level with point 12′ in the elevation, and the true length diagonal taken up to the point level with 11′. This precaution must be observed with all the zigzag lines which terminate on the joint line from 8 to 14.

BREECHES WITH PREDETERMINED JOINT LINES

The examples shown in Fig. 131 represent breeches pieces in which the form of the joint line JL may be made to any desired shape. In

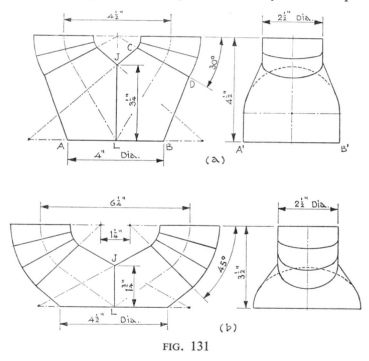

FIG. 131

Fig. 131(a) it will be seen that the joint line, as shown in the side elevation, takes the form of a semicircle in the upper portion with sides falling vertically to the base diameter $A'B'$. In Fig. 131(b) the form of the joint line is an ellipse with the minor axis in the vertical position, and the major axis equal to the diameter of the base.

146   Sheet Metal Drawing and Pattern Development

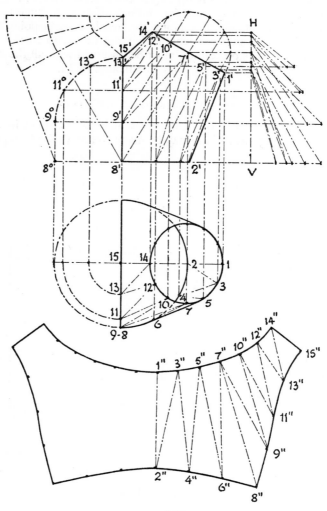

FIG. 132

Development by Triangulation 147

The pattern for the breeches piece illustrated in Fig. 131(a) is shown developed in Fig. 132. In general the procedure is similar to that of the previous example in Fig. 130. The difference lies chiefly in connexion with the shape of the joint line from 8 to 15. In this case one half of the true shape of the joint line is drawn on the left-hand side in the elevation, as from 8^0 to 15'. The quadrant 9^0 to 15' is divided into three equal parts, while the vertical portion from 8^0 to 9^0 forms a fourth division not necessarily equal to the other three. Now, points 11^0 and 13^0 are projected vertically to the horizontal centreline in the plan, and from the centrepoint 15 are swung round to points 11 and 13 on the joint line.

The plan is now ready for dividing into triangles, beginning at the outside line from point 1 to point 2, and following the usual zigzag pattern from one point to the next between the top and bottom edges, then passing on round the surface to points 14 and 15 on the opposite side, as shown in Fig. 132. In this example it should be noted that an extra point occurs on the joint line, which is best seen in the elevation at point 9^0. This point lies vertically above point 8^0, which means that the single point 8 also serves for point 9 in the plan.

The procedure for developing the pattern is precisely the same as that already described in connexion with the example shown in Fig. 130. The only difference in the process is the inclusion of the extra triangle 7"8"9", which arises from the extra point 9 on the joint line. The true spacings 8"9", 9"11", 11"13" and 13"15" are taken direct from the true shape of the joint line in the elevation.

BREECHES WITH SHALLOW ELLIPTICAL JOINT LINE

The setting out for the development of the pattern for one limb of the breeches piece shown in Fig. 131(b), is given in Fig. 133, and is again similar to that of Fig. 130. The difference lies in the shape of the joint line, which in this case is elliptical with the minor axis in the vertical position. One half of the true shape is plotted on the left-hand side of the joint line in the elevation, Fig. 133. The curve is divided into three equal parts and the points projected vertically to the horizontal centreline in the plan. Then, from the centrepoint 14, the points on the centreline are swung round to obtain the points 10 and 12 on the joint line in the plan.

The plan is now divided into triangles by a zigzag line which alternates between the top and bottom edges from points 1 to 14. The development of the pattern is again obtained by the same procedure as that described in connexion with Fig. 132. The chief points to bear in mind are that all the lines forming the zigzag which pass from top to bottom or bottom to top must be triangulated against their respective vertical heights on *VH* in order to obtain their true lengths, and that the true spacings for the top edge must

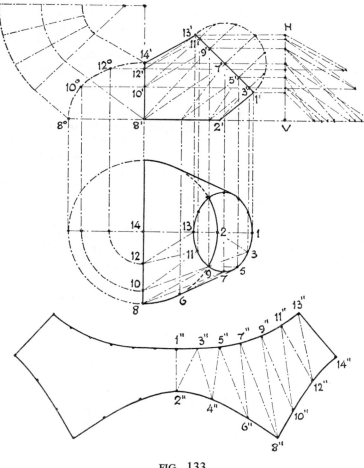

FIG. 133

be taken from the semicircle in the elevation and not from the ellipse in the plan. The true spacings for the bottom edge are taken from the quadrant 2 to 8 in the plan, and from the elliptical curve 8^0 to $14'$ in the elevation.

Development by Triangulation

BREECHES PIECE WITH ODD LIMBS

The example shown in Fig. 134 represents a branch piece in which the two limbs taper to different diameters, though they terminate on the same base their centre lines meet on the base and they have a common joint line between them. In this case it is obvious that one pattern will not serve for both limbs, and that both sides must be triangulated for the development of their respective patterns. The shape of the joint line on *xy*, Fig. 134, is a semi-ellipse.

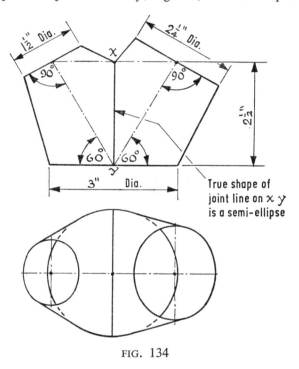

FIG. 134

Referring to Fig. 135, one quarter of the ellipse is drawn on the left-hand side of the joint line in the elevation, using 7'13' as half the major axis and 7'7⁰ as half the minor axis. This elliptical curve represents the shape of the joint line on 7'13' turned through 90° in order to show its true form. The curve is now divided into three parts appropriately spaced, though not necessarily equal, and the two intermediate points 9⁰ and 11⁰ are projected horizontally back to the vertical joint line to obtain points 9' and 11'. The position of the points 9 and 11 in the plan are now determined by taking

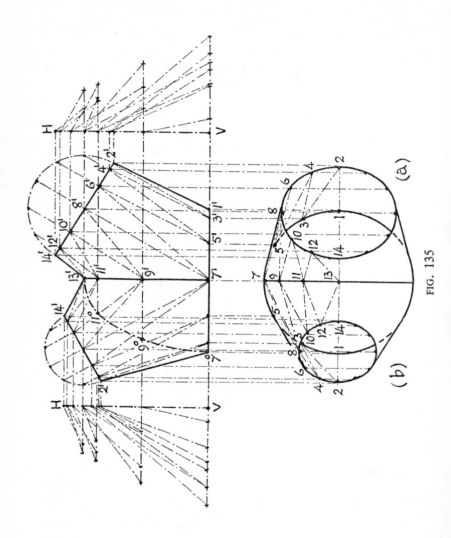

FIG. 135

Development by Triangulation

the horizontal distances 9°9′ and 11°11′ from the elevation and marking them off from the centrepoint 13 in the plan. This method of obtaining these points on the joint line in the plan differs somewhat from the method described in connexion with Figs. 130, 132 and 133, but the principle is basically the same and the same result is arrived at.

The circular top edges of the two branches are dropped into the plan as ellipses by the usual method of describing a semicircle on the diameter at the top of each, dividing it into six equal parts, projecting the points back to the diameter, and dropping them into the plan to obtain the respective ellipses.

The plan is now ready for dividing the two surfaces into triangles. Taking first the larger right-hand limb, the quadrant 1 to 7 on the base is divided into three equal parts, as at 1, 3, 5 and 7. The ellipse has already been divided into six parts (though not equal ones) in the process of plotting it from the elevation. Beginning at the outside line 1,2, the corresponding points around the top and bottom edges are joined, as 3 to 4, 5 to 6, 7 to 8, 9 to 10, 11 to 12, and 13 to 14. The diagonals are next drawn from 2 to 3, 4 to 5, and so on, thus forming a zigzag line round the surface of the limb.

Next, the vertical height line VH is drawn in the elevation, and all the points on the top edge 2′ to 14′ are projected horizontally to it. Also, the points 9′, 11′ and 13′ are projected through it to serve as base lines.

The development of the patterns is shown in Fig. 136, that for the larger limb at (*a*) and for the smaller limb at (*b*). The first line in the pattern for the larger limb may be taken direct from the elevation, as from 1′ to 2′, as that line is already a true length; this distance may then be marked off in any convenient position, as from 1″ to 2″. For the next line, the plan length 2 to 3 is taken and marked off at right angles to the vertical height along the base line level with V. The true length diagonal is taken up to the point level 2′, and from point 2″ in the pattern an arc is drawn through point 3″. To complete the first triangle, the true distance 1 to 3 is taken direct from the plan, and from point 1″ in the pattern an arc is drawn cutting the previous arc in point 3″.

For the next triangle, the plan length 3 to 4 is taken and marked off from V along the base line. The true length diagonal is taken up to the point level with 4′, and from point 3″ in the pattern an arc is drawn through point 4″. Next, the true spacing 2″4″ is required. This spacing is not obtained from the ellipse in the plan, but is taken from the semicircle in the elevation, and from point 2″ in the pattern an arc is drawn cutting the previous arc in point 4″. This completes the second triangle.

The four triangles which follow, up to the line 7″8″, are obtained by the same procedure. Then for the next or seventh triangle, the plan length 8 to 9 is taken and marked off at right angles to the vertical height, but this time along the line level with point 9′ in the elevation. The true length diagonal is taken up to the point level with 8′, and from point 8″ in the pattern an arc is drawn through point 9″. Next, the true spacing 7″ to 9″ is obtained direct from the

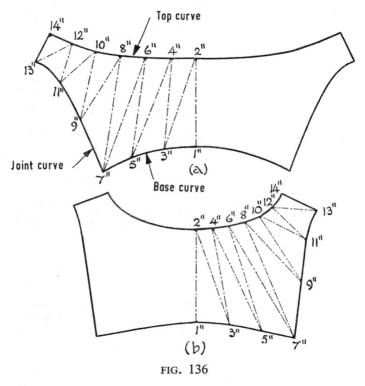

FIG. 136

true curve of the joint line in the elevation, as from 7⁰ to 9⁰, and from point 7″ in the pattern an arc is drawn cutting the previous arc in point 9″.

In determining the remaining triangles in the pattern it is essential that the plan lengths 9,10, 10,11, 11,12 and 12,13 should be marked off along their respective base lines level with points 9′, 11′ and 13′ in the elevation. Their true lengths are then obtained by taking the respective diagonals up to the corresponding points level with 10′, 12′ and 14′ on the top edge.

Development by Triangulation

The method of developing the pattern for the smaller or left-hand limb is precisely the same as that for the larger limb. The same system of numbering the points is used, those on the joint line being common to both sides. The vertical height line is placed appropriately on the left-hand side, and the secondary base lines through points 9′, 11′ and 13′ are again common to both sides. The procedure for developing the pattern should readily be followed from the description given for the right-hand limb. The developed pattern is shown in Fig. 136(b).

BREECHES PIECE ON SQUARE BASE

The breeches piece shown in Fig. 137 represents two branches terminating on a square base which could fit on a square outlet

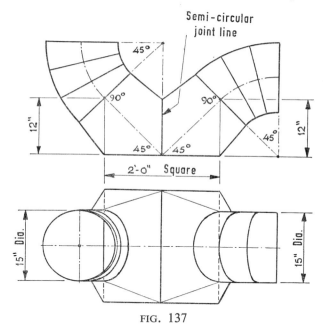

FIG. 137

connexion or to a square trunking. The joint between the two branches is semicircular, with its diameter across the middle of the square. The development of the pattern is shown in Fig. 138.

In setting out the problem for pattern development, only a half plan of the right-hand limb is drawn below the elevation. The semi-ellipse in the plan is obtained in the usual way by drawing a semicircle on the diameter in the elevation, dividing it into six equal parts, projecting the points back to the diameter, and dropping them vertically into the plan so that the widths of the ellipse may be marked off equal to the corresponding widths taken from the semicircle. Since the joint line between the limbs is semicircular, its true

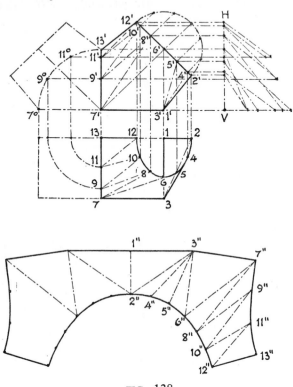

FIG. 138

shape, which is required for the pattern development, may be represented by the quadrant drawn on the left-hand side in the elevation.

Before dividing the plan into triangles, the positions of the two points 9 and 11 on the joint line must be determined. To do this, the quadrant in the elevation is divided into three equal parts, giving points 9^0 and 11^0 on the curve. These two points are dropped vertically onto the horizontal centreline in the plan, and, using point 13 as centre, they are then swung round to the vertical centreline, or joint line, giving points 9 and 11. Then, arranging the seam

Development by Triangulation

to occur on the top of the limb from 12 to 13, the surface is divided into triangles as shown in the illustration from points 1 to 13.

Points 9 and 11 will occur in the elevation at points 9′ and 11′, which are obtained by drawing horizontal lines from points 9^0 and 11^0 to cut the vertical joint line in 9′ and 11′. These horizontal lines are further extended to cut the vertical height line VH, in order to be used as secondary base lines in the process of development.

To begin the pattern, the line from 1 to 2 in the plan may be taken and triangulated against its vertical height; alternatively, as it is horizontal in the plan, its true length may be taken direct from the elevation from 1′ to 2′. This distance is then marked off in any convenient position for the first line 1″ to 2″ in the pattern. Next, the plan length from 2 to 3 is taken and marked off along the base line from point V. The true length diagonal is taken up to the point level with 2′ in the elevation, and from point 2″ in the pattern an arc is drawn through point 3″. The true distance from 1 to 3 is then taken direct from the plan, and from point 1″ in the pattern an arc is drawn cutting the previous arc in point 3″. This completes the first triangle.

For the second triangle the plan length 3 to 4 is taken and marked off from V at right angles to the vertical height. The true length diagonal is taken up to the point level with 4′, and from point 3″ in the pattern an arc is drawn through point 4″. Next, the true spacing 2″ to 4″ is obtained, not from the ellipse in the plan, but from the semicircle on the top edge in the pattern, and from point 2″ in the pattern an arc is drawn cutting the previous arc in point 4″.

The next two triangles, 3″4″5″ and 3″5″6″, are obtained in a similar manner by triangulating the plan lengths 3 to 5 and 3 to 6 against their appropriate vertical heights and taking their respective true lengths up to points level with points 5′ and 6′ in the elevation. The true spacings 4″5″ and 5″6″ will again be taken from the semicircle on the top edge in the elevation.

For the fifth triangle, the plan length 6 to 7 is taken and marked off from V at right angles to the vertical height. The true length diagonal is taken up to the point level with 6′ in the elevation, and from point 6″ in the pattern an arc is drawn through point 7″. Next, the true length line 3 to 7 is taken direct from the plan, and from point 3″ in the pattern an arc is drawn cutting the previous arc in point 7″.

From here, the remainder of the pattern should offer no difficulty provided that the correct vertical heights are used in relation to the points on the top edge in the elevation, and the true spacings for that edge in the pattern are taken from the semicircle on the top edge. Also, the true spacings for the joint edge 7″9″11″13″ are taken from the quadrant in the elevation. Since the branch piece is symmetrical about the horizontal centreline, the pattern will be symmetrical about the centreline 1″2″ and the other half of the pattern is repeated on the opposite side of that line.

THREE-WAY BREECHES PIECE

The three-way breeches piece illustrated in Fig. 139 is a further example in which the joint line between the branches is predetermined and specified as an ellipse, in this case with semi-axes of 12 in.

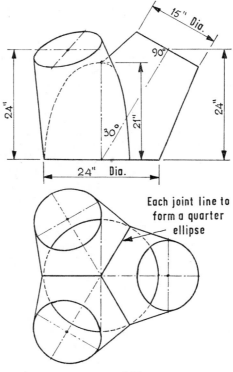

FIG. 139

and 21 in. The setting out for development, which is shown in Fig. 140, offers some variation on the previous examples of two-way branches, although the basic principles are the same.

Since the branches are all equal, only one is set out for development in Fig. 140, and only one half of that is divided up for triangu-

Development by Triangulation

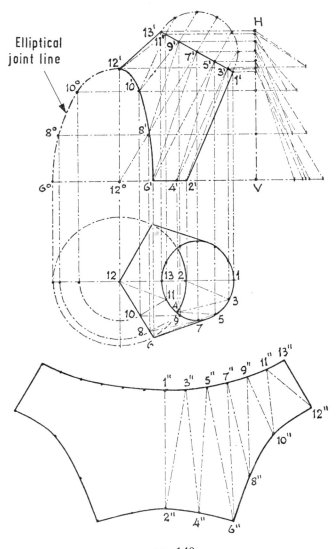

FIG. 140

lation. The joint line between the branches occurs as from 6 to 12 in the plan, and lies at an angle of 60° to the horizontal centreline.

Before dividing the surface into triangles, the two points 8 and 10 must be determined on the joint line in the plan. To do this, the quarter ellipse 6^0 to $12'$ is drawn in the elevation on the semi-axes $12^0 6^0$ and $12^0 12'$ by any method of drawing ellipses or ovals. The curve is then divided into three parts, equal or otherwise, giving the points 8^0 and 10^0. These points are then dropped vertically to the horizontal centreline in the plan, and, using point 12 as centre, they are swung round to the joint line, thus determining points 8 and 10. The surface in the plan is then divided into triangles as shown in Fig. 140, arranging for the seam to occur on the top, as from 12 to 13.

To determine the shape of the joint line in the elevation, points 6, 8 and 10 are projected vertically upwards from the plan to meet horizontal lines from point 6^0 on the base line and points 8^0 and 10^0 on the elliptical curve, thereby locating points $6'$, $8'$ and $10'$. A curve now drawn from point $6'$ through $8'$ and $10'$ to $12'$ will represent the form of the joint line as seen in the elevation.

To develop the pattern the first line from 1 to 2 is taken from the plan and triangulated against its vertical height, or as this line is horizontal in the plan its true length may be taken direct from the elevation as from $1'$ to $2'$; then the line $1''$ to $2''$ is marked off in any convenient position to begin the pattern. Next, the plan length 2 to 3 is taken and marked off at right angles to the vertical height VH. The true length diagonal is taken up to the point level with $3'$ in the elevation, and from point $2''$ in the pattern an arc is drawn through point $3''$. Now, the true spacing $1''3''$ required for the pattern is taken, not from the plan, but from the semicircle on the top edge in the elevation, and from point $1''$ in the pattern an arc is drawn cutting the previous arc in point $3''$.

For the second triangle the plan length 3 to 4 is taken and marked off from point V at right angles to the vertical height. The true length diagonal is taken up to the point level with $3'$, and from point $3''$ in the pattern an arc is drawn through point $4''$. Next, the true spacing 2,4 is taken direct from the plan, and from point $2''$ in the pattern an arc is drawn cutting the previous arc in point $4''$.

The next three triangles, $3''4''5''$, $4''5''6''$ and $5''6''7''$, are obtained in a similar manner by triangulating the plan lengths 4,5, 5,6 and 6,7 against their appropriate vertical heights, taking the true spacings for the top edge from the semicircle in the elevation, and the true spacing for the bottom edge from 4 to 6 in the plan.

For the next triangle the plan length 7 to 8 is taken and marked off at right angles to its vertical height, but this time along a base

Development by Triangulation

line level with point 8' in the elevation. The true length diagonal is then taken up to the point level with 7' on the top edge, and from point 7" in the patern an arc is drawn through point 8". The true spacing 6"8" in the pattern is obtained this time from the corresponding true spacing 6^08^0 on the elliptical joint line in the elevation, and from point 6" in the pattern an arc is drawn cutting the previous arc in point 8".

The remainder of the pattern should not be difficult to follow from this point. The chief precautions are to triangulate the plan lengths against their appropriate vertical heights, take the true spacings for the top edge from the semicircle in the elevation, and take the remaining true spaces for the bottom edge from the elliptical joint line in the elevation.

Junction Pieces

The type of branch piece illustrated in Fig. 141 is of different practical construction from the breeches type so far considered. Instead of the curved joint line between the branches, the middle portion consists of flat triangular pieces which are bent along the

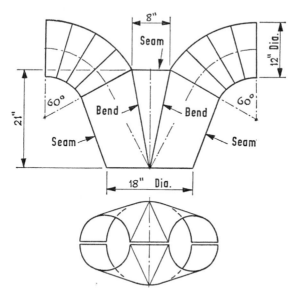

FIG. 141. *Plan of junction piece only, showing construction in two halves*

sides, and have a short joint line along the top of the triangle between the two halves of the branch piece. The main seams are disposed at the outsides of the branches as shown in Fig. 141. In this illustration the length of the central seam at the top of the triangles is somewhat exaggerated for the purpose of making the construction more clear. In actual practice the length of this seam, which represents the width across the top of the triangles between the branches, need not be more than 1–1½ in. even in large-sized branch pieces, since the closer the branches are together the more efficient the branch piece normally is.

This type of branch piece will be referred to as a "junction piece," to differentiate it from breeches piece, though both serve the same purpose. As has been said, the chief difference lies in the practical construction.

TWO-WAY JUNCTION PIECE

The development of the pattern is shown in Fig. 142, and consists of straightforward triangulation. The most important point to observe in the development of this type of branch piece is the method of dividing the surface into triangles. First, it should be noted that on each side of the flat triangle whose apex is shown at point 2 in the plan, the construction is a quarter of a cone. The cone is in the inverted position with its apex at point 2 and its base forming a quarter of the circular top of the branch. Therefore, the radial lines from point 2 at the bottom to points 3, 4, 5 and 6 at the top all lie on the surface of the conical portion.

The remainder of the body of the branch is *not* conical. It will be observed that the quarter circle at the bottom, numbered 2,7,9,11, is joined by the zigzag line to the corresponding quarter circle at the top, numbered 6,8,10,12. Since these two quarter circles are not in parallel planes, the portion of the body between the quadrants cannot form a portion of a cone.

The geometry of this junction piece should be carefully studied, as nearly all branch pieces of this type follow this basic principle of construction, including some of complex design. Haphazard methods of dividing the surface into triangles can result in totally incorrect patterns even though the principles of triangulation are followed in the development. When the body of the junction piece is correctly analysed and divided up accordingly, the pattern development should offer no difficulty. The complexity of junction piece construction will be appreciated by an inspection of the three-way branch piece shown in Fig. 244 (p. 314) and the six-way piece shown in Fig. 270 (p. 357). It will also be appreciated that this

Development by Triangulation 161

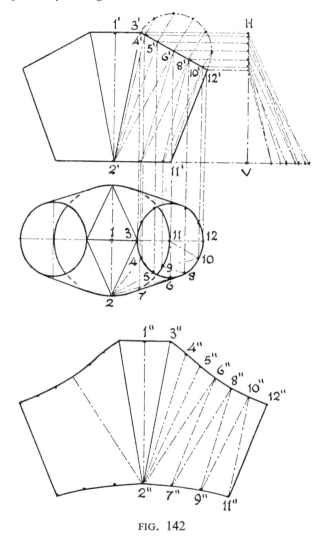

FIG. 142

method is more flexible and allows of greater complexity of construction than the breeches-piece type.

To develop the pattern for the junction piece shown in Fig. 142, the first plan length from 1 to 2, which is at the centre of the flat triangle, is taken and marked off along the base line at right angles to the vertical height *VH*. The true length diagonal is taken up to the top point level with 1' in the elevation, and the first line in the pattern, 1″ to 2″, is marked off in any convenient position. Next, the plan length from 2 to 3 is taken and marked off at right angles to the vertical height. The true length diagonal is taken up to the top point level with 3', and from point 2″ in the pattern an arc is drawn through point 3″. Now, the true distance 1,3 is taken from the plan, and from point 1″ in the pattern an arc is drawn cutting the previous arc in point 3″.

For the second triangle the plan length 2,4 is taken and marked off at right angles to the vertical height. The true length diagonal is taken, this time up to the point level with 4' in the elevation, and from point 2″ in the pattern an arc is drawn through point 4″. Next, the true spacing 3″4″ in the pattern is taken, not from the plan, but from the semicircle on the top edge in the elevation. Then from point 3″ in the pattern an arc is drawn cutting the previous arc in point 4″.

The remainder of the pattern should not be difficult to follow from this point. The chief precautions to observe are that the true length diagonals should be taken up to the appropriate points on the vertical height line, and that the true spacings for the top edge should be taken from the semicircle on that edge in the elevation.

SQUARE-TO-TWO CIRCLES JUNCTION PIECE

The junction piece shown in Fig. 143 is a typical branch piece used in ductwork, which transforms from a square at one end to two

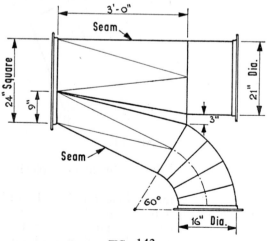

FIG. 143

Development by Triangulation 163

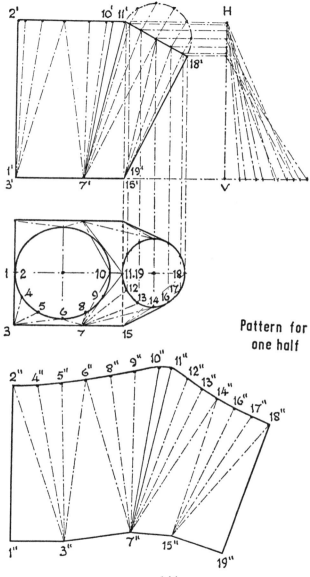

FIG. 144

unequal circles at the other. The development of the pattern is shown in Fig. 144, in which the junction piece is turned through 90° to stand with the square end as the base. The seams are located on the outsides of the junction piece with the flat triangles in the middle. In accordance with junction piece construction it will be seen that the apex of the flat triangle at point 7 is also the apex of the two quarter cones, one on each side of the triangle. This may be better seen in the elevation, where the apex lies at point 7'.

The first line in the pattern may be taken direct from the elevation from 1' to 2', and marked off in any convenient position as from 1" to 2" to begin the pattern. Next, the plan length from 2 to 3 is taken and marked off at right angles to the vertical height VH. The true length diagonal is taken up to the top point level with 2', and from point 2" in the pattern an arc is drawn through point 3". Now, the true distance from 1 to 3 is taken direct from the plan, and from point 1" in the pattern an arc is drawn cutting the previous arc in point 3".

For the second triangle the plan length 3,4 is taken and marked off at right angles to the vertical height VH. The true length diagonal is taken up to the top, and from point 3" in the pattern an arc is drawn through point 4". Next, the true spacing from 2 to 4 is taken direct from the plan, and from point 2" in the pattern an arc is drawn cutting the previous arc in point 4".

From this point onwards the development of the pattern should not be difficult to follow since the process is straightforward triangulation, and further instruction would be largely a repetition of terms. It should be observed, however, that the points on the left-hand circle from point 1 to point 10 and across the triangle to point 11 all take the full vertical height on VH, while the points on the smaller circle from 11 to 18 move consecutively downwards in accordance with their positions on the slope or angle of the top edge in the elevation. Also, the true spacings for the top edge in the pattern must be taken from the semicircle on the top edge in the elevation and *not* from the ellipse in the plan.

7 Intersections by Cutting Planes

When two geometrical bodies intersect or interpenetrate each other, the line of intersection must be determined in the plan or elevation or both before the patterns for the two bodies can be developed. Such bodies as cylinders, prisms, pyramids, cones and spheres are common objects which occur as intersecting bodies in sheet metal work.

There are two important methods by which lines of intersection can be determined. One is the method of cutting planes. The other is the method of auxiliary projection.

The present chapter deals with the method of cutting planes. A cutting plane, as used in this connexion, is an imaginary flat plane

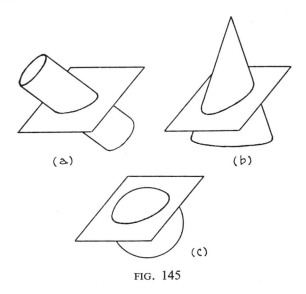

FIG. 145

which cuts through both bodies, thereby enabling one or two points to be located on the joint line. Three of four cutting planes will generally be sufficient to enable the position and shape of the joint line to be determined.

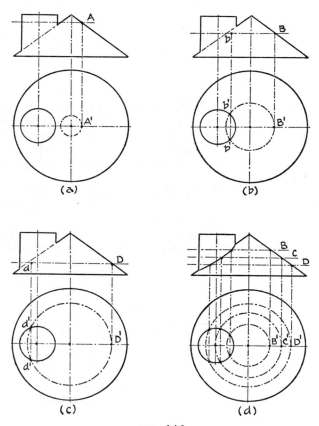

FIG. 146

The basic idea of a cutting plane is illustrated in Fig. 145, in which a cylinder is cut by a plane at (*a*), a cone at (*b*), and a sphere at (*c*).

The application of the method in determining the joint line is shown in Fig. 146, in which a vertical cylinder penetrates a cone. In Fig. 146(*a*), a horizontal cutting plane is taken through the point *A*, which lies just above the highest point of intersection between the two bodies. The plan of the cone at the position of the cutting

Intersections by Cutting Planes

plane is a circle as shown at A'. The plan of the cylinder at the same plane is also a circle as shown at the side of the conic circle. Note that the two circles do not cut each other.

In Fig. 146(b) the cylinder and cone are exactly the same size and intersect each other in the same position. A horizontal cutting plane is taken at B, a point little lower than A and within the limiting top and bottom points of intersection. Again, the plan of the cone at the position of the cutting plane is a circle as shown at B'. The plan of the cylinder at the same plane is again a circle of the same size and in the same position as that in Fig. 146(a). Note that in the present case, the two circles cut each other at points b', b'. If, now, these points are projected vertically upwards to the cutting plane in the elevation, a point will be obtained at b which lies on the joint line, or line of intersection.

In Fig. 146(c) a horizontal cutting plane is taken at point D, which is further towards the bottom than point B but still within the limiting points of intersection. Once again, the plan of the cone at the position of the cutting plane is a circle as shown at D', and the conic circle cuts the cylinder circle, this time at points d', d'. Furthermore, these points projected vertically upwards will give point d on the cutting plane in the elevation. This point also lies on the line of intersection.

From the foregoing illustrations it will readily be seen that if this process be repeated with a number of cutting planes, taken in any position within the top and bottom limits of the intersection, sufficient points will be obtained in the elevation to enable the line of intersection to be drawn in.

In Fig. 146(d) three cutting planes are taken at positions B, C and D. The corresponding circles are drawn in the plan at B', C' and D'. The points at which these circles cut the circle of the cylinder are projected vertically upwards to the corresponding cutting planes in the elevation. Thus, sufficient points are afforded through which to draw the line of intersection or the joint line in the elevation. Having obtained the position and shape of the joint line, the patterns for the cone and cylinder may readily be obtained by radial line and parallel line development.

RIGHT CONIC CONNEXION TO CONICAL COVER

The right conic connexion to a conical cover shown in Fig. 147 represents a typical example which calls for the determination of the joint line before the pattern can be developed. The joint line is obtained by the method of cutting planes as illustrated in Fig. 148.

First, the elevation and plan are drawn of the two cones in their relative positions. Then, a number of cutting planes, as represented by the horizontal lines through A', B' and C' in the elevation, are drawn between the top and bottom points of intersection. The points A', B' and C' are now projected vertically to the horizontal centreline in the plan to locate points A, B and C, which provide radial points through which the plan circles of the conical cover may be drawn.

Next, the other extremities of the cutting planes in the elevation, a', b' and c', are projected vertically to the horizontal centreline

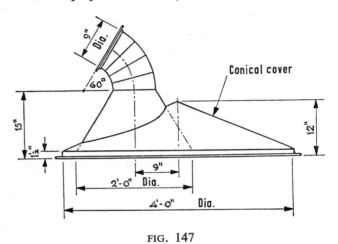

FIG. 147

in the plan to locate points a, b and c, which provide radial points through which the plan circles of the conic connexion may be drawn.

The two sets of circles in the plan cut each other at points d, e and f, and these points are now projected vertically upwards to the corresponding cutting planes in the elevation to locate points d', e' and f'. These points lie on the joint line in the elevation, which may now be drawn in as shown in the Figure.

Having obtained the joint line, the pattern for the conic connexion may be developed in either of two different ways, though both are based on the same principles. The connexion is of right conic construction, with apex at O'. As with right conic development, any points on the joint line which are made use of in the development must first be projected to the side of the cone. In this case the points d', e' and f', which occur on the cutting planes, are already, in effect, projected to the side of the cone to points a', b' and c'. These

Intersections by Cutting Planes

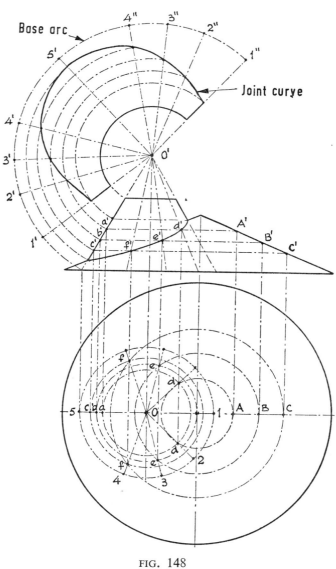

FIG. 148

points may be projected into the pattern together with the other points down that side of the cone as shown in Fig. 148.

The chief difference from the usual right conic procedure lies in the spacings around the base arc in the pattern. To obtain these, the radial lines on the cone must pass through points d', e' and f' in the elevation, and through points d, e and f in the plan. Therefore, from point O in the plan, which represents the apex, lines are drawn through points d, e and f to the circle which represents the base of the cone, thereby locating points 1, 2, 3, 4 and 5. It will be seen that the divisions on the circle made by these points are not equal.

Now, the unequal spacings from 1 to 5 are taken and marked off along the base arc in the pattern, as from $1'$ to $5'$, and then repeated in the reverse order from $5'$ to $1''$. From these points on the base arc, radial lines are drawn to the apex O'. Where the radial lines cross the arcs from the side of the cone, points will be afforded through which to draw the joint curve, as shown in Fig. 148.

The Alternative Method The alternative method of developing the pattern for the connexion depends simply on first obtaining the joint line by cutting planes, and then, ignoring the points thus obtained on the joint line, proceeding with the development by the usual right conic method. This procedure is illustrated in Fig. 149.

Three cutting planes are taken at A', B' and C', and the joint line obtained by the same procedure as in the previous example. Then, ignoring the construction lines used in the cutting plane process, a semicircle is described on the base of the cone and divided into the usual six equal parts. The points on the semicircle are numbered from 1 to 7. From these points, lines are projected upwards perpendicular to the base, and from the points on the base, radial lines are drawn to the apex O'. The points where the radial lines cross the joint line are now projected horizontally to the outside of the cone. Note that these points on the joint line are not the same as those obtained in the process of determining the joint line by cutting planes.

From this point the process of developing the pattern should be readily followed from the illustration since it is similar to examples already dealt with in the chapter on radial line development. The first of these methods has the advantage of using fewer lines, while the second may result in a more accurately shaped pattern.

OBLIQUE CONIC CONNEXION TO CONICAL COVER

The conical cover and connecting piece shown in Fig. 150 present a similar case to that of Fig. 147, except that in Fig. 150 the inter-

Intersections by Cutting Planes

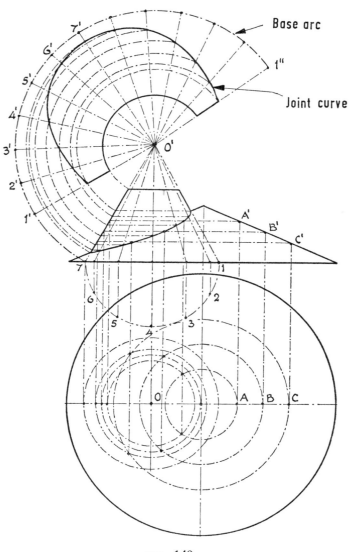

FIG. 149

secting cone is an oblique instead of a right cone. The determination of the joint line and the development of the pattern are shown in Fig. 151.

Three cutting planes are taken in the positions of A', B' and C' in the elevation. These points are dropped into the plan, or half-plan as shown, to locate the points A, B and C. There is one point of difference from the previous example which should be noted. As the intersecting cone is oblique, its centreline is not vertical, but inclines at an angle from the centre of its base to its apex at O'.

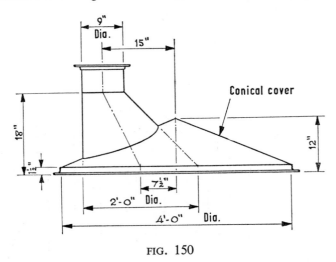

FIG. 150

This means that, although its cross-section at the cutting planes is circular, the centre of each circle in the plan occurs vertically below its position in the elevation. Therefore care is needed in locating the correct centres when describing the oblique conic circles in the plan. Then the points where the corresponding circles cut each other are projected vertically upwards to their respective cutting planes to obtain the points on the joint line.

Having now determined the shape of the joint line in the elevation, the pattern may be obtained by the usual methods of oblique conic development. Thus, ignoring the construction lines used in the process of cutting planes, the semicircle is described on the base in the elevation and divided into the usual six equal parts which are shown numbered from 1 to 7. The points on the semicircle are then projected vertically upwards to the base line, and from the points on the base, elevation lines are drawn to the apex O'.

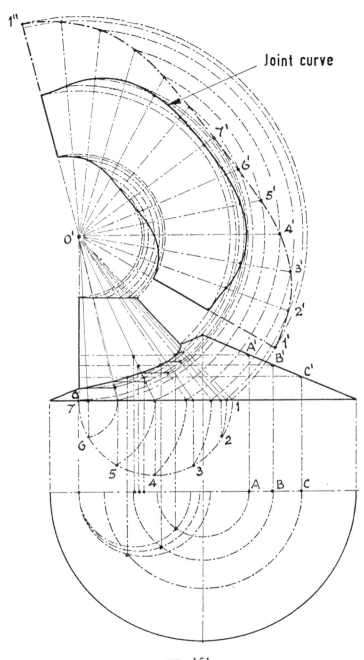

FIG. 151

Now, the apex O' falls vertically to the point O at the left-hand extremity of the base. Using point O as centre, arcs are drawn from the points on the semicircle to the base. Again, from these points on the base, lines are drawn to the apex O'. These lines are true length lines and are shown dotted to differentiate them from the corresponding elevation lines.

Next, using the apex O' as centre, the true length lines are swung into the pattern from the base line and also from the points where they cross the top edge of the cone. Then one of the equal divisions is taken from the semicircle and, beginning on the outside arc, spacings are stepped over from one arc to the next as shown in the

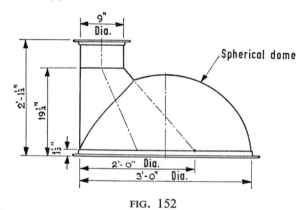

FIG. 152

Figure from 1' to 7', and then back to the outside arc at 1". From those points radial lines are drawn to the apex O'.

The next stage in the process of development is to determine the true distances from the apex O' to the various points on the joint line. It will be noted that two sets of lines cross the joint line: these are the elevation lines and true length lines. It should also be noted that one pair of these lines, an elevation line and a true length line, emanate from each point on the semicircle. Therefore each true length line represents the true length of its corresponding elevation line. Thus, a horizontal line drawn from the point where an elevation line crosses the joint line to the corresponding true length line, will give the true length from the apex O' to that point on the joint line.

Therefore from each of the points where elevation lines cross the joint line, horizontal lines are drawn to the corresponding true length lines. These true length lines are then swung into the pattern, and where they cross the radial lines already drawn, points are afforded

Intersections by Cutting Planes

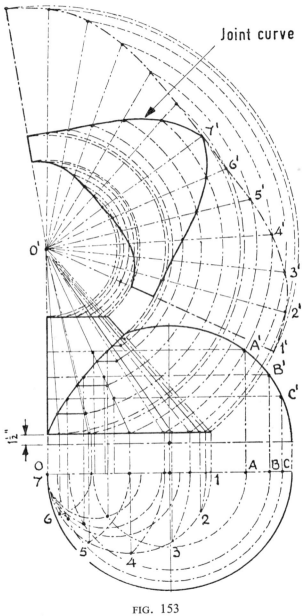

FIG. 153

through which to draw the joint curve in the pattern as shown in Fig. 151. The curve representing the top edge may now be drawn to complete the pattern.

OBLIQUE CONIC CONNEXION TO SPHERICAL DOME

The example shown in Fig. 152 represents an oblique conical connexion to a spherical dome. The process of obtaining the joint line by the method of cutting planes is basically the same as that described in connexion with the previous example, since cutting planes through the sphere and cone also present circles in the plan.

The determination of the joint line and the development of the pattern are shown in Fig. 153. In this example a half plan is used for the cutting planes, and the half plan used for the development of the pattern is superimposed on that of the cutting planes. Apart from this difference in the setting out, the procedure in the course of solution is precisely the same as that given in connexion with Figs. 149 and 151.

RIGHT CONE INTERSECTED BY RECTANGULAR PIPE

The right cone intersected by a rectangular pipe illustrated in Fig. 154 is a further example in which the line of intersection may be

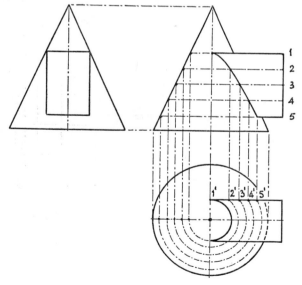

FIG. 154

Intersections by Cutting Planes

determined by the method of cutting planes. The shape of the curve made by the intersection of the vertical sides of the pipe is hyperbolic, while the curves made by the top and bottom of the pipe are arcs of circles.

In this example five cutting planes are taken, including the top and bottom faces of the pipe, as at 1, 2, 3, 4 and 5 in the elevation. From these points horizontal lines are drawn to the other side of the cone, to represent the cutting planes. From the points on the other side of the cone, lines are dropped vertically to the horizontal centreline in the plan. From the centre of the circle in the plan, circles or arcs are then drawn to cut the vertical sides of the pipe, as at 1', 2', 3', 4' and 5'. These points are now projected vertically upwards to the corresponding cutting planes in the elevation, thereby locating points through which the hyperbolic curve may be drawn.

A VENTILATOR HEAD

The ventilator head illustrated in Fig. 155 represents an application of the cone and rectangular pipe intersection. The determination of the line of intersection and the development of the patterns are shown in Fig. 156.

Five cutting planes are taken in the positions shown at points 1, 2, 3, 4 and 5 in the elevation, and the shape of the joint line is obtained as described in the previous example.

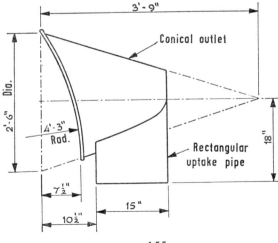

FIG. 155

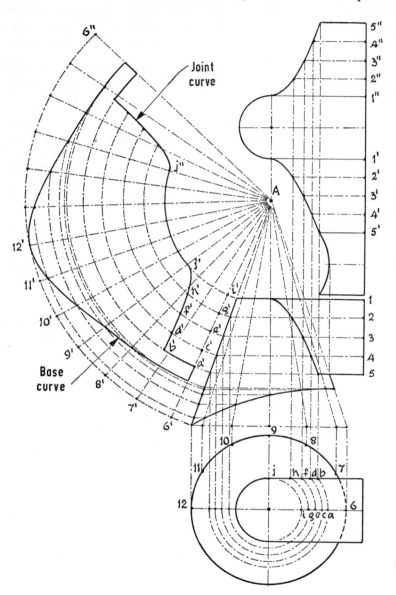

FIG. 156

Intersections by Cutting Planes

To develop the pattern for the conical part, the semicircle in the plan, which represents the base of the cone, is divided into six equal parts, as from 6 to 12, and the points projected vertically upwards to the base of the cone in the elevation. From the points on the base of the cone, radial lines are drawn to the apex A. Then from the points where the radial lines cross the curved bottom edge, horizontal lines are drawn to the outside of the cone.

Next, all the points on the side of the cone, including those of the cutting planes, are swung into the pattern. One of the equal divisions is now taken from the semicircle in the plan, and twelve spacings marked off along the outside arc in the pattern, as from 6' to 12' and onwards to 6". From these points radial lines are drawn to the apex A, and where the radial lines cross the arcs drawn from the side of the cone, points are afforded through which to draw the base curve in the pattern.

Now, to obtain the joint curve in the pattern, the lengths of the curves, a to b, c to d, e to f, g to h, and i to j, are taken from the plan and marked off from the outside radial lines along their respective arcs in the pattern, as from a' to b', c' to d', e' to f', g' to h', and i' to j'. From point j' the curve of the pattern lies along the inside arc until it meets the corresponding joint j'' on the other side. The final part of the curve is a duplicate of the first part from a' to j'.

The pattern for the rectangular pipe is obtained by parallel line development and straightforward "unrolling," and it should be readily followed from the illustration.

CONICAL SPOUT ON DOMED VESSEL

The illustration in Fig. 157 represents a cylindrical vessel, the top part of which is dome shaped with the centre part flat. The spout is a portion of a right cone superimposed on the flat and curved part of the top. The determination of the joint line and the development of the pattern are shown in Fig. 158.

In the solution of this problem only two cutting planes have been taken, although any number of cutting planes could be used in order to obtain more points on the line of intersection. The cutting planes are horizontal planes passing through the points marked A and B, which may be located in any desired position. These planes cut through both the top of the vessel at A and B, and the conical spout at a and b.

The plans of the cutting planes on the vessel are circles, as shown in the half plan at A' and B'. The plans of the cutting planes on the conical spout are also circles, as shown in the half plan at a' and b'.

The points where the two corresponding circles intersect, as at c' and d', are projected vertically upwards to c and d on the corresponding cutting planes in the elevation. These points represent two positions located on the line of intersection. The top point e where the line of intersection leaves the quadrant curve of the dome, is situated vertically above the centre of radius of the quadrant. The portion of the intersection from e to f lies on the flat centre part of the top of the vessel.

Once the line of intersection is drawn in, the remainder of the problem consists of simple right conic development. The semicircle

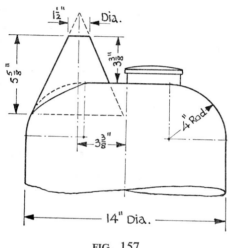

FIG. 157

is drawn on the base of the cone, divided into six equal parts and the points numbered from 1 to 7. The points are then projected vertically upwards to the base of the cone, and from the points on the base of the cone, radial lines are drawn to the apex O. The points where the radial lines cross the joint line are projected horizontally to the side of the cone, and with apex O as centre, the points on the side of the cone are swung into the pattern. Next, one of the equal divisions is taken from the semicircle and twelve spacings marked off around the base arc in the pattern, as from $1'$ to $7'$ and onwards to $1''$. From the points on the base arc radial lines are drawn to the apex O.

Before the base curve in the pattern is drawn in, extra points x' and x'' are required on the base arc to locate the exact position of points e' and e'' in the pattern. To obtain these points, a line is

Intersections by Cutting Planes 181

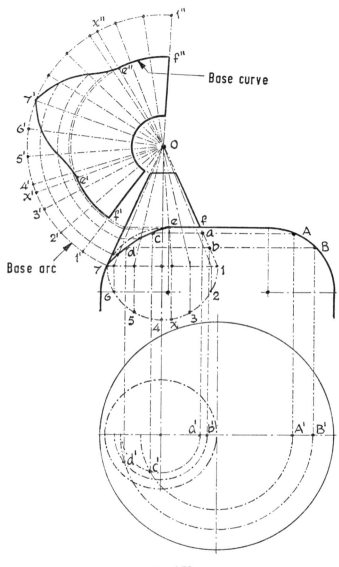

FIG. 158

drawn from the apex *O* through point *e* to the base of the cone in the elevation. From the point on the base of the cone a line is dropped perpendicularly to the semicircle to locate the point *x*. Then, the distance from 3 to *x* is taken and marked off along the base arc in the pattern from 3' to *x'*. A similar point *x"* is then located on the opposite side of the pattern. From these two points extra radial lines are drawn to the apex *O*. The two extra lines then effect the location of points *e'* and *e"* on the inside arc. The base curve and the remainder of the pattern is now drawn as shown in Fig. 158.

CONICAL COVER WITH SIGHT SCREEN HOLDER

The illustration in Fig. 159 represents a right conical cover intersected by a rectangular prism with semicircular sides. This example is one in which the application of cutting planes is the most convenient method of determining the line of intersection, though an

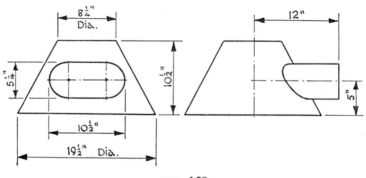

FIG. 159

alternative method based on radial lines on the cone might be used. The determination of the joint line, the development of the pattern for the prism, and the contour of the hole in the conical cover are shown in Fig. 160.

In this case a plan is dropped from the side elevation, and a half shape of the cross-section of the prism set off from the end *GH*. In the elevation the semicircle on *EF* represents the shape at each side of the prism. This semicircle is divided into six equal parts, and from the points on the semicircle horizontal lines are drawn right through the prism and the conical cover. These horizontal lines are taken as the cutting planes through both bodies. The plan of the cone at each of the cutting planes will be a circle of diameter corresponding to its position. Thus, the plan circles of the cutting planes

Intersections by Cutting Planes

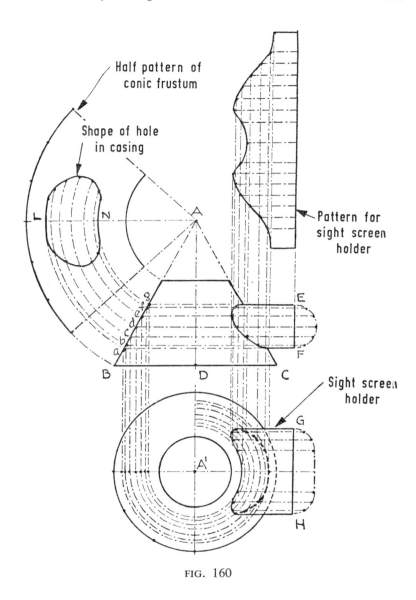

FIG. 160

may be obtained by dropping the points a, b, c, d, e, f and g vertically downwards to the horizontal centreline in the plan. Then from centre A' the plan circles are drawn from the points obtained on the centreline. In Fig. 160 it will be noted that only three-quarters of the circles are drawn, but these are sufficient for the purpose of locating the joint line.

Next, the two quadrants at each end of GH in the plan are each divided into three equal parts and the points projected horizontally to cut the circles on the cone. The lines projected from GH represent the plan positions of the cutting planes through EF, and the careful plotting of the points of intersection with the corresponding conic circles gives sufficient points through which to draw the joint line in the plan. The points obtained on the joint line are now projected vertically upwards to the corresponding cutting planes in the elevation to give sufficient points through which to draw the joint line in that view.

The pattern for the prism, or sight screen holder, is "unrolled" vertically from the elevation. The shape of the hole in the conical cover or casing is obtained by describing arcs from the points a, b, c, d, e, f and g, using the apex A as centre. Then a centreline LN is drawn through the arcs to the apex A in any convenient position. Next, the lengths of the arcs on both sides of this centreline are obtained from the plan by taking the distances on either side of the horizontal centreline to the edge of the hole and marking them off on the corresponding arcs above and below LN. The pattern for one half of the full conic frustum is then completed by marking off one-quarter of the circumference of the base along the outside arc on each side of the centreline LN in the pattern, and joining the extremities to the apex A.

8 Auxiliary and Double Projections

The examples dealt with in this chapter are not so much concerned with the development of the pattern as with the methods of determining the shape of the joint line in an appropriate view from which the pattern may be developed. The methods adopted are those of auxiliary and double projection applied with the object of obtaining the required view.

It often occurs, particularly in pipe- and ductwork, that the layout of a scheme of work given in normal orthographic plan and elevation does not present certain sections in sufficient detail to enable pattern developments to be carried out without seeking further views which involve auxiliary projections. There are certain basic principles underlying the process of obtaining auxiliary projections which might well be introduced before dealing with the problems which follow. It is possible to obtain a series of several auxiliary projections, as illustrated in Figs. 162 and 163, though usually one or two projections are sufficient to solve most of the problems which require this method of solution. Double projections are those in which a second projection is made from the first projection in order to obtain the necessary view for the development of the pattern. In obtaining double or multiple projections the idea to bear in mind is that the object, in each successive view, is turned or rolled through 90° in the direction of the projection.

Projections on Straight Line

In Fig. 161 a single line is dealt with to illustrate perhaps the most important basic rule. Complex problems merely constitute a multiplicity of this rule. The elevation, Fig. 161, is simply a straight line MN standing in the vertical position. The plan is therefore represented by a single point. If, now, it is desired to obtain a view looking in the direction of arrow A, then the line is laid over or turned through an angle of 90° in the direction of the arrow. Since the line is standing vertically, and is therefore at right angles to the plane of the paper in the plan, its full height will be seen in the direction of the arrow A. The first projection is then represented by the straight line MN shown at (1), which is equal to the full length of the line. Furthermore, it is lying flat or horizontal in the plane of the paper.

The second projection is taken in the direction of the arrow B, and the two extremities, M and N, are projected downwards to the position M^0N^0 in the second projection shown at (2). Since the line

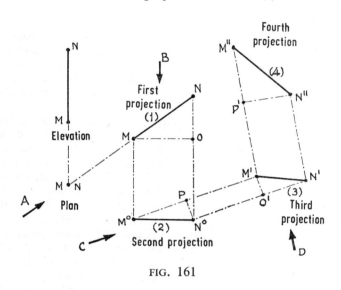

FIG. 161

MN in the first projection is lying flat on the paper, in the second projection it will be seen at right angles to the direction of the projection. It will be noted that the line M^0N^0 in the second projection is shorter than that in the first projection, and is therefore not the full or true length of the line. In fact, in the second projection, from point M^0 the line rises from the plane of the paper to the point N^0, which lies at a height above M^0 equal to ON in the first projection. This is an important point to be borne in mind for the next or third projection.

The third projection is taken in the direction of the arrow C, and the two extremities, M^0 and N^0, are projected in that direction to the position $M'N'$ in the third projection shown at (3). A base line, $M'O'$, is drawn in any convenient position at right angles to the direction of the lines of the projection. Now, to obtain the point

Auxiliary and Double Projections

N', the height ON is taken from the first projection and marked off from point O' in the third projection to give $O'N'$ in that view. The reason for this is that the distance ON in the first projection, being turned through 90° into the second projection, is at right angles to the plane of the paper. In the next move it is again laid over at 90° to give the distance $O'N'$ in the third projection.

The next point in the process is an important one, and one which may not be easily visualized. In order to obtain the third projection, the object in the second projection, in this case a single line, is revolved or turned through 90°. This means that point M' is now uppermost, and is higher than point N' by a distance equal to PM^0 in the second projection. The line PN^0 is drawn at right angles to the projection lines.

The fourth projection is now made in the direction of the arrow D. The two points M' and N' are projected to the positions M'' and N''. The base line $P'N''$ is drawn in any convenient position at right angles to the projection lines. Now to obtain point M'', the distance PM^0 is taken from the second projection and marked off along the projection line from P' in the fourth projection. Then the line $M''N''$ is obtained by joining these two points. Again, the reason for this move is that the distance PM^0 in the second projection, being turned through 90° into the third projection, is at right angles to the plane of the paper and represents the difference in height between M' and N'. It is again laid over at 90° to give the height $P'M''$ in the fourth projection.

It will be seen from this example that any number of projections can be made by following the principle of rolling or turning the object through 90° in any desired direction. The important rule to remember is that all the heights marked from a base line in the new projection are taken from the second view back in the series. Thus, $O'N'$ in the third projection is taken from ON in the first projection, and $P'M''$ is taken from PM^0 in the second projection.

Projections on Square Prism

Following up the example shown in Fig. 161, a square prism is dealt with in Fig. 162 on the same basic principles. The ordinary elevation and plan are shown on the left-hand side, and the first projection is made in the direction of the arrow A. The height of the prism in the direction of the projection is the same as the height in the elevation, that is, MN in the first projection is equal to MN in the elevation. This is consistent with the single line example in Fig. 161, and also with the rule that "heights" or distances marked from a base line in any projection are the same as the corresponding heights in the second view back.

The second projection is made in the direction of the arrow *B*. All the points on the ends of the prism in the first projection are projected downwards into the second projection. The next move should be carefully noted. In the plan a datum line is drawn at right

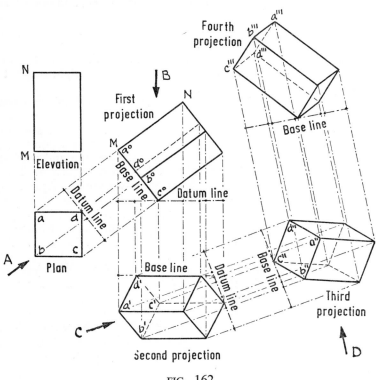

FIG. 162

angles to the direction of the first projection, and touches the nearest point in the plan on the side of the first projection. Also a base line is drawn in any convenient position across and at right angles to the projection lines in the second projection. Now, to obtain the second projection, the distances are taken from the datum line in the plan to each of the four corner points *a*, *b*, *c* and *d*, and marked off from the base line on the corresponding lines in the second projection. This determines the parallelogram $a'b'c'd'$. The points at the other end of the prism have the same distances marked off from the base

Auxiliary and Double Projections

line to give a similar parallelogram. The two parallelograms are joined by lines representing the corners of the prism.

The third projection is made in the direction of the arrow C. Again, all the points on both ends of the prism in the second projection are projected forward into the third projection. Also, in the first projection a datum line is drawn at right angles to the projection lines between the first and second projections. This datum line touches the lowest point in the first projection. A base line is also drawn across and at right angles to the projection lines in the third projection. Now, to obtain the third projection, the distances are taken from the datum line in the first projection to each of the four corner points a^0, b^0, c^0 and d^0, and marked off from the base line on the corresponding lines in the third projection, thereby obtaining the parallelogram $a''b''c''d''$. The heights of the points at the other end of the prism are also taken from the datum line in the first projection and marked off from the base line on the corresponding lines in the third projection, thereby locating the parallelogram which represents the other end of the prism. The corresponding corners of the two parallelograms are now joined by straight lines, thus completing the view of the prism in the third projection.

The fourth projection is made in the direction of the arrow, D and again all the corner points on both ends of the prism in the third projection are projected upwards, not vertically, into the fourth projection. A datum line is now drawn in the second projection at right angles to the projection lines between the second and third projections. As in the previous cases, the datum line touches the point in the second projection on the side nearest the third projection. A base line is now drawn across and at right angles to the projection lines in the fourth projection. The process of obtaining the fourth projection is now similar to those of the second and third, inasmuch as the distances are taken from the datum line in the second projection to each of the corner points a', b', c' and d', and marked off from the base line on the corresponding lines in the fourth projection. For example, the distance in the second projection from the datum line to point a' is taken; the projection line from a' is followed along to a'' in the third projection, and from a'' the projection line is further traced upwards into the fourth projection; the distance taken from the second projection is then marked off from the base line to give point a'''. This is repeated with all the points on both ends of the prism to obtain the two parallelograms in the fourth projection. The corresponding corners of the parallelograms are now joined by the straight lines which complete the view in the fourth projection as shown in Fig. 162.

The importance of this series of projections is to emphasize the rule that the heights or distances marked from the base line in any projection are taken from the datum line or its equivalent in the second view back. Thus, in this series, the distances from the base line to the various points in the fourth projection are taken from the datum line in the second projection. Similarly, the distances from the base line in the third projection are taken from the datum line in the first projection, and the distances from the base line in the second projection are taken from the datum line in the plan.

Projections on Right Cone

The series of projections shown in Fig. 163 are made with a right cone as the basic model. The directions and angles in which the

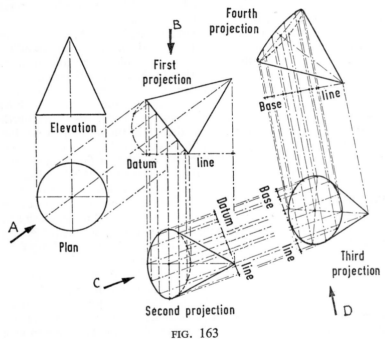

FIG. 163

projections are made are similar to those in Fig. 162, and in general the methods of obtaining the views are also similar to those given in connexion with the previous Figure.

The second projection should readily be obtained from the first projection by the usual method of drawing the semicircle on the

Auxiliary and Double Projections

base of the cone, as shown, and dropping the base into the second projection as an ellipse. The apex is then dropped from the first projection to the centreline in the second projection; the centreline is drawn through the centre of the ellipse at right angles to the projection lines.

The ellipse in the third projection is obtained by projecting the points on the ellipse in the second projection into the third projection, and the distances, taken from the datum line in the first projection, are marked off from the base line on the corresponding lines in the third projection. Also, the height of the apex in the first projection is taken from the datum line and marked off on the corresponding line from the base line in the third projection.

Similarly, to obtain the fourth projection, the points around the ellipse in the third projection are projected forward into the fourth projection and the base line marked off in any convenient position. Then from the datum line in the second projection, the distances to the points on the ellipse are taken and marked off on the corresponding lines from the base line in the fourth projection, thus obtaining the points on the ellipse in that view. Now since the apex lies on the datum line in the second projection, it also lies on the base line in the fourth projection, as will be seen in Fig. 163.

If the principles underlying the projections illustrated in Figs. 161, 162 and 163 are carefully followed and understood, no real difficulty should be experienced in following the solutions of the succeeding examples set out in this chapter. It should be clearly understood, however, that in most cases the development of the pattern is a relatively simple matter compared with the methods and principles which must be applied in obtaining the correct view from which the pattern may be developed. Moreover, it is advisable to complete each view in the process whether or not it is the one which will finally be required for the development of the pattern, for facility in the determination of joint lines only comes with practice.

CHUTE FROM SCREW CONVEYOR

The chute from the screw conveyor shown in Fig. 164 has an off-centre lower limb which requires an auxiliary projection from which the pattern is developed. The plan is of the chute only, and shows it to be of cylindrical construction with the front edge of the lower portion cut off at 45° to its centreline. The projection and pattern development are shown in Fig. 165.

The auxiliary projection is made from the plan at right angles to the centreline AB. A ground line $D''E''$ is drawn across and at right angles to the projection lines, and the point A'' is marked as the centre of the ellipse. Also, in the elevation, a similar ground line, $D'E'$, is drawn through the centre A' of the ellipse.

Now, to obtain the centrelines of the chute in the projection, the height $B'C'$ is taken from the elevation and marked off in the projection from B'' to C''. Also, the height $C'F'$ is taken from the

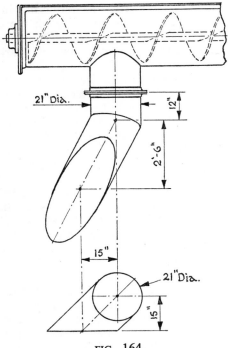

FIG. 164

elevation and marked off in the projection from C'' to F''. The centrelines $A''C''F''$ now represent their true lengths and are lying flat in the plane of the paper.

Next, the diameter of the chute is drawn through F'' at right angles to the centreline $C''F''$. The outsides of the elbow are now drawn with the joint line through C'' as shown in the Figure. The elliptical edge has yet to be determined.

A semicircle is drawn on the diameter, divided into six equal parts, and the points thereon are projected back to the diameter and onwards through the joint line to the circle in the plan. This divides the circle into twelve equal parts similar to those on the

Auxiliary and Double Projections

semicircle in the projection. From the points on the circle, lines are drawn parallel to the centreline AB to meet the edge of the chute in the plan. From the points on the edge of the chute, lines are drawn back to the projection, and where these meet a corresponding series of lines drawn from the joint line parallel to the

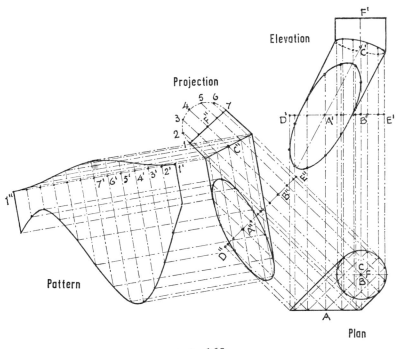

FIG. 165

centreline $A''C''$, points are afforded through which to draw the ellipse as shown in the illustration.

To obtain the ellipse in the elevation, lines are projected upwards from the points on the bottom edge in the plan, and then, from the projection, heights and depths are taken above and below the ground line $D''E''$ to the various points on the ellipse, and marked off on the corresponding lines above and below the ground line $D'E'$ in the elevation. Points are thereby afforded through which to draw the ellipse in the elevation.

The elliptical joint line in the elevation is obtained in a similar manner. The points around the circle in the plan are projected upwards into the elevation. The heights from the ground line $D''E''$

194 Sheet Metal Drawing and Pattern Development

in the projection up to the joint line are then taken and marked off on the corresponding lines from the ground line $D'E'$ in the elevation, thereby obtaining points through which the elliptical joint curve is drawn.

The pattern for the lower part of the chute is shown developed or unrolled from the projection, and as the procedure is straight forward parallel line development, there should be no difficulty at this stage in following it from the illustration.

INTERSECTION OF TWO SQUARE PIPES

To obtain the development of the pattern for the branch pipe shown in Fig. 166 an auxiliary projection is made from the plan at right

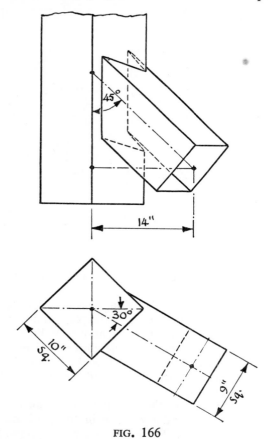

FIG. 166

Auxiliary and Double Projections

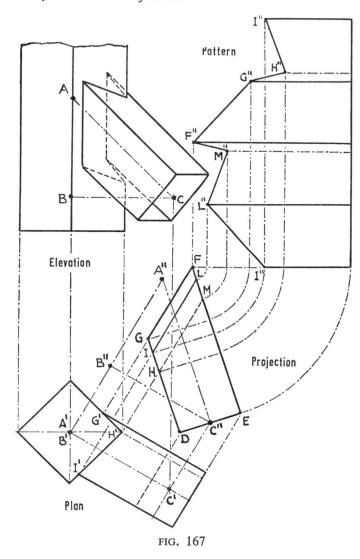

FIG. 167

angles to the centreline of the branch pipe. The projection, from which the pattern for the branch pipe is unrolled, is shown in Fig. 167.

First, the centrepoints A' and C' are projected at right angles to the centreline $A'C'$, and a base line, $B''C''$, is drawn parallel to $A'C'$ in any convenient position. Next, the height BA is taken from the elevation and marked off from B'' to A'' in the projection. Then the line $A''C''$ represents the true length of the centreline of the branch pipe and is lying flat in the plane of the paper. The end of the pipe, DE, is now drawn through C'' at right angles to $A''C''$, and also the sides DG and EF are drawn parallel to $A''C''$. To complete the auxiliary projection, the points G', H' and I' are projected from the plan into the projection to cross the two sides as shown in the Figure.

The unrolling of the pattern is a straightforward parallel line development and should be readily followed from the illustration. One point, however, is worthy of note. The angles $L''M''F''$ and $G''H''I''$ in the pattern are not right angles. It may at first be thought that as those faces of the pattern have to fit against the corner of a square pipe, as at $G'H'I'$ in the plan, the angles in the pattern must be 90°; however, the slope of the branch pipe in the elevation results in the corner angles in the pattern becoming acute angles.

SQUARE PIPE ELBOW ON SQUARE MAIN

The illustration in Fig. 168 represents a front elevation and a side elevation of a square pipe elbow intersecting a square main pipe at an angle. An auxiliary projection and the development of the patterns are shown in Fig. 169.

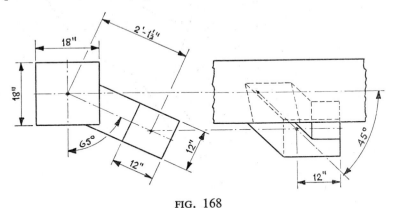

FIG. 168

The projection is made from the left-hand elevation at right angles to the centreline AC, and a base line $A''B''$ is drawn across and at right angles to the projection lines. Then the distance $B''C''$ is obtained from the second view back, that is, from B' to C' in the right-hand elevation, and marked off in the projection from B'' to

C''. Also, the distance $C''D''$ is obtained from the right-hand elevation, as from C' to D', and marked off along the same projection line from C'' to D''. The auxiliary projection is now completed by drawing the semi-width of the pipe on each side of the centrelines $A''C''$ and $C''D''$, and projecting the points E, F and G from the

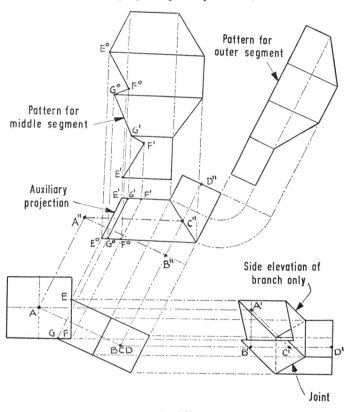

FIG. 169

left-hand elevation into the projection to meet the sides of the pipe elbow as shown in the Figure.

Since the centrelines of the elbow, $A''C''$ and $C''D''$ are lying flat in the plane of the paper, the projection is therefore suitably placed for unrolling or developing the patterns. The middle segment, which is connected to the main pipe, is unrolled at right angles to the centreline $A''C''$, and the procedure may readily be followed from the illustration. The pattern for the outer portion of the elbow on the centreline $C''D''$ would normally be unrolled at right angles to $C''D''$, but in the present case, Fig. 169, it is turned through 90° from the normal position for the convenience of compact setting out.

SQUARE PIPE ELBOW ON CYLINDRICAL PIPE

The example shown in Fig. 170 represents, in plan and elevation, a square pipe elbow connected to a cylindrical pipe. In order to obtain

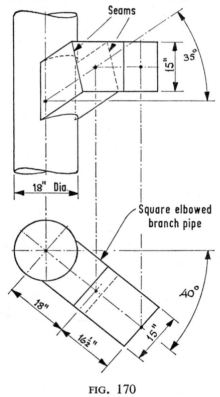

FIG. 170

the necessary view of the pipe and its centrelines lying flat in the plane of the paper, an auxiliary projection is made from the plan. The projection and development of the patterns are shown in Fig. 171.

From the plan, the points A, B, C and D are projected at right angles to the centreline, and a base line $A''B''$ is drawn in any

Auxiliary and Double Projections

convenient position. Then the height $B''C''$ is obtained from the second view back (in this case from the elevation, as shown from B' to C'), and marked off from B'' to C'' in the projection. Since the centreline $C'D'$ of the outer segment is horizontal in the elevation, that portion is drawn parallel to $A''B''$ in the projection to give

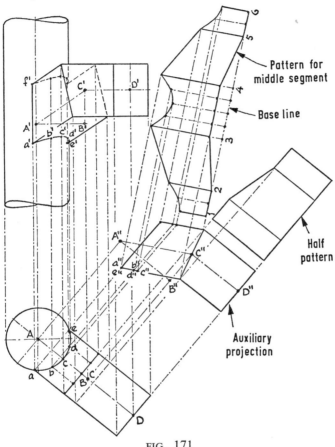

FIG. 171

$C''D''$, which is equal in length to CD in the plan. Now, the centrelines $A''C''$ and $C''D''$ in the projection are lying flat in the plane of the paper, which is the first condition towards obtaining the view from which the patterns may be unrolled.

As the branch pipe is of square cross-section, its semi-width is drawn on each side of and parallel to the centrelines $A''C''$ and

$C''D''$ as shown in the Figure. Next, as the lower end of the elbow intersects the cylindrical pipe, extra points are made around the curve in the plan by dividing the width of the pipe into four equal parts and projecting them onto the curve to obtain the points a, b, c, d and e. These points are now projected into the auxiliary projection to cross the middle segment of the elbow in points a'', b'', c'', d'' and e''. It will be seen that points a'' and b'' also stand for d'' and e''.

Now, to obtain the elliptical curve of the joint line $a'b'c'd'e'$ in the elevation, the points a, b, c, d and e are projected vertically upwards from the plan, and then the distances from the base line $A''B''$ in the auxiliary projection are taken downwards to a'', b'', c'', d'' and e'', and marked downwards on the corresponding lines from the base line $A'B'$ in the elevation. Points are thereby afforded through which the elliptical joint curve is drawn. The corresponding curve above, on the opposite face of the pipe, is obtained by taking the distance $a'f'$ and marking this vertically upwards from each of the points a', b', c', d' and e'.

The pattern for the middle segment is unrolled from the auxiliary projection at right angles to the centreline $A''C''$. The seam is arranged to occur in the middle of the short side on the top of the segment. Therefore the widths of the two sides and bottom faces are marked off along the base line and one half of the width of the top is marked off at each end to bring the seam in the middle of the top. The two end spaces, 1 to 2 and 5 to 6, are divided into two equal parts, and the middle space, 3 to 4, is divided into four equal parts. From all the points along the base, lines are drawn at right angles to meet the series of lines projected from the various points on the middle segment in the projection.

The shape of the pattern should not be difficult to follow from the illustration since it is based on straightforward parallel line development. A half pattern for the outer segment is also shown unrolled at right angles to the centreline $C''D''$.

CYLINDRICAL PIPE ELBOW ON CYLINDRICAL MAIN

The cylindrical pipe elbow represented in Fig. 172 shows the middle segment entering the cylindrical main pipe at 45° in the elevation and 30° in the plan. As the centrelines of the elbow in the plan present a straight line, the solution of the problem depends on a simple auxiliary projection. The projection and development of the patterns are shown in Fig. 173.

The auxiliary projection is made from the plan at right angles to the centrelines ACD. The points A, C and D are first projected

Auxiliary and Double Projections

forward, and a ground line $A''C''D''$ drawn across and at right angles to them. The ground line may be drawn in any convenient position. The height $A''B''$ is next determined by taking the distance $A'B'$ from the elevation and marking it off from A'' in the projection. Then, by joining B'' and C'', the centrelines $B''C''$ and $C''D''$ are obtained lying flat in the plane of the paper. The patterns may then be unrolled from the projection.

First, the pipe diameters are drawn parallel to the centrelines, and the straight joint line drawn through C'' as shown in the Figure.

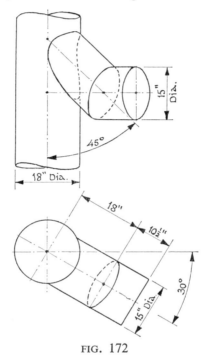

FIG. 172

Next, to determine the joint line with the cylindrical main pipe, semicircles are drawn on the ends of the branch elbow, as at D and D''. The semicircles are each divided into six equal parts, and from the points thereon, lines are drawn parallel to the centrelines of the pipe. In the plan, these lines pass straight through to the circle representing the main pipe. In the projection, the lines from the semicircle are drawn to the joint line through point C'', and from there are drawn parallel to the centreline $C''B''$. Now, from the points in the plan where the parallel lines meet the main pipe circle, lines are drawn into the projection to meet the corresponding parallel lines in that view. Points are thereby afforded through which to draw the joint curve in the projection.

The pattern for the middle section on the centreline $B''C''$ may now be unrolled as shown in the Figure. It will be noted that the pattern for the end section of the elbow is developed above that for

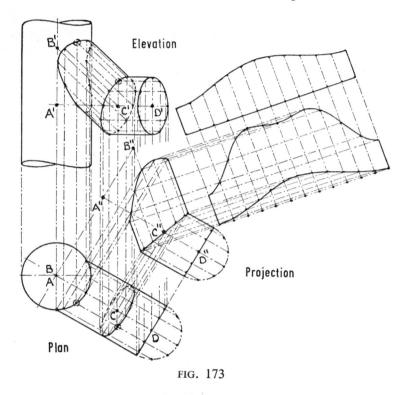

FIG. 173

the middle section, and that the same division lines are extended upwards for use in that pattern.

It may be observed that detailed directions for the developments are not given. At this stage it is assumed that the reader is familiar with the ordinary methods of development, and attention is therefore concentrated mainly on methods of projection, the determination of joint lines, and the completion of curves and ellipses in the various views involved in the setting out of the problem. It is perhaps as much a question of draughtsmanship as of pattern development.

Auxiliary and Double Projections 203

Even so, in most cases in this chapter a joint line *must* be determined before the pattern can be developed.

To complete the plan view, the ellipse around point C may readily be obtained by projecting the points back from the joint line at C'' to meet the corresponding lines in the plan.

Next, the two ellipses in the elevation may be obtained from the plan. First, the ellipse around the point D' in the elevation may be obtained by projecting the points along the edge D from the plan into the elevation and marking off the appropriate widths above and below the horizontal centreline through point D' thus obtaining the necessary points through which to draw the ellipse.

The second ellipse around point C' in the elevation is similarly obtained, in this case by projecting the points around the ellipse C from the plan into the elevation. Then lines are drawn horizontally from the points on the first ellipse to meet the corresponding lines drawn from the plan, thereby affording points through which to draw the second ellipse around C' as shown in the Figure.

Now, in the elevation, the shape of the joint line between the elbow and the main cylinder is obtained by projecting the points from the plan circle into the elevation to meet a series of corresponding lines drawn parallel to the centreline $B'C'$ from the points on the ellipse around C'.

These lines form closed circuits between the plan and elevation, as may be seen by following the lines between the points with the small rings round them as illustrated in Fig. 173. The point on the ellipse in the plan corresponds to, or is the same point as, that with the ring round it on the ellipse in the elevation. These points are joined by the vertical line between them. Again, the point on the ellipse in the plan lies on the parallel line which meets the main circle in the point with the ring round it. Now, the line drawn from the point on the circle into the elevation meets the line drawn from the ellipse in the elevation at the point shown with the small ring round it. Thus the four points, two in the plan and two in the elevation, are joined by the four lines and form a closed quadrilateral. By careful observation, it will be seen that the lines drawn from the circle in the plan meet their corresponding lines drawn from the ellipse in the elevation, thereby affording points through which the shape of the joint line is drawn, as illustrated in Fig. 173.

ANOTHER CYLINDRICAL ELBOW ON CYLINDRICAL MAIN

Although the cylindrical elbow illustrated in Fig. 174 is somewhat similar to that shown in Fig. 172, it has one important difference which alters the method of approach to the determination of the joint lines and the pattern development. It will be seen that in the plan view, Fig. 172, the centrelines of the two parts of the elbow form a straight line, from which a single auxiliary projection is sufficient for the solution of the problem. In the case of the example shown in Fig. 174, two projections are necessary before the joint lines and the development of the patterns can be obtained. The projections and patterns are shown in Fig. 175.

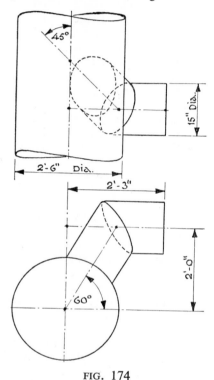

FIG. 174

The primary object of the projections is to obtain a view in which both centrelines of the elbow are lying flat in the plane of the paper. The first step towards this is to observe that the centreline CE of the outer limb of the elbow is already lying flat in the plane of the paper both in the plan and in the elevation; also, that the centreline BC of the middle section rises from the plane of the paper from C to B in the plan and elevation. Next, a side view is projected from the plan such that the centreline CE of the outer limb stands perpendicularly to the plane of the paper. Thus the outer limb then

Auxiliary and Double Projections

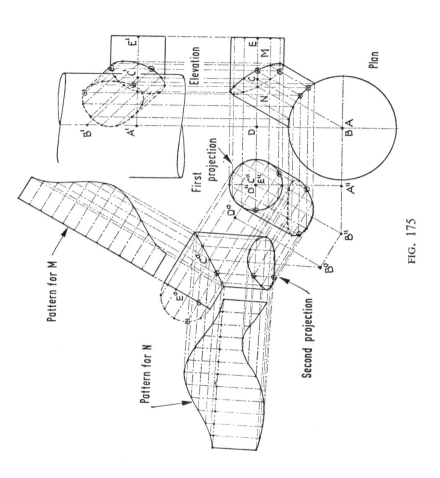

FIG. 175

presents a circle in the projection, the centre of which is lettered D'', C'', E''.

Now, the ground line or datum line $D''A''$ is drawn through the centre D'' of the circle and made parallel to DA in the plan. Next, from the centrepoint A of the main pipe circle in the plan, a line is drawn into the projection at right angles to DA, and the distance $A''B''$ is marked off equal to $A'B'$ from the elevation. The line $B''C''$ then represents the centreline of the middle segment in the projection.

The next step is to obtain a second projection at right angles to the centreline $B''C''$. These two points are therefore projected at right angles to $B''C''$ into the second projection, and a ground line B^0D^0 is drawn in any convenient position parallel to $B''C''$.

Now, in the second projection, the point C^0 lies above point D^0 at a distance equal to DC taken from the plan. Also, the point E^0 lies above C^0 at a distance equal to CE taken from the plan. Thus the full height D^0E^0 is equal to the full distance DE in the plan. The centrelines B^0C^0 and C^0E^0 represent their true lengths and are now lying flat in the plane of the paper. The pipe diameters are next drawn parallel to the centrelines, and the top edge of the outer limb is drawn through E^0 at right angles to the centreline. The joint line at the elbow is also drawn through point C^0 as shown in the Figure.

The shape of the joint line at the bottom edge of the middle segment has now to be determined in the first and second projections. First, the semicircle is drawn on the top edge at E^0, divided into six equal parts, and lines therefrom are drawn parallel to the centreline of the limb to the joint line through C^0. These lines are continued onwards to the circle in the first projection, thus dividing the circle into a number of equal parts similar to those around the semicircle on the top edge in the second projection.

The next step is to project the points on the circle back into the plan to meet the outer edge of the limb at E. The ellipse around the centrepoint C in the plan is now obtained by taking the distances from the edge at E^0 to the joint line through C^0 in the second projection, and marking these distances backwards from the edge at E on the corresponding lines in the plan. Points are thereby afforded through which to draw the ellipse around point C.

From the points on the ellipse around C, lines are now drawn parallel to the centreline BC to meet the main circle as shown in the plan. The points thus obtained on the main circle are projected back to the first projection to meet a corresponding series of lines drawn parallel to the centreline $B''C''$ from the points on the circle around C''. Points are thereby afforded through which to draw the shape of the joint line in the first projection.

Auxiliary and Double Projections

To obtain the shape of the bottom joint line in the second projection, the points around the joint line in the first projection are projected into the second projection to meet a corresponding series of lines drawn parallel to the centreline B^0C^0 from the points along the joint line through C^0. Where the two sets of lines meet, points are afforded through which to draw the shape of the joint line as shown in the Figure.

A summary of this procedure, starting from the semicircle in the second projection, is outlined by the series of points with small circles round them. The point on the semicircle is projected back to the edge E^0, and onwards through the joint line C^0 to the circle in the first projection, where it gives two points on the circle, each of which is ringed with a small circle. Both of these points are projected to the edge at E in the plan, and on both lines a distance is marked backwards equal to that between the edge E^0 and the joint line C^0 in the second projection. This locates the two ringed points on the ellipse in the plan. These two points are now projected to the circumference of the main circle in the plan. From there, two lines are drawn into the first projection to meet the pair of lines drawn from the corresponding points on the circle in that view. Finally, the two ringed points thereby obtained on the joint line in the first projection are projected into the second projection to meet the line drawn from the corresponding ringed point on the joint line at C^0. Thus the two ringed points are obtained on the bottom joint line in the second projection.

Having now obtained the bottom joint line in the second projection, the patterns for both parts of the elbow may be unrolled from that view as shown in Fig. 175. Detailed directions for the unrolling of the patterns are not given, as the process from this point is one of straightforward parallel line development.

To complete the example, the two joint lines in the elevation may now be determined. The points on the ellipse in the plan are projected into the elevation. The points around the ellipse in the elevation are then obtained from the second view back, which in this case is the first projection, by taking the distances to left and right of the centreline $A''C''$ to each of the points on the circle, and marking these distances on the corresponding lines above and below the centreline $A'C'$ in the elevation.

This process may be followed more readily by observing the series of ringed points. From the first projection the distance from the centreline $A''C''$ to the ringed point on the left is taken and marked off above the centreline $A'C'$ in the elevation on the corresponding line. The corresponding line is that which may be followed from the ringed point in the first projection, back to the plan, and upwards to the elevation. A similar process may be followed from the other ringed point on the right of the centreline $A''C''$ in the first projection, by taking its distance from the centreline and marking it off below the centreline $A'C'$ on the corresponding line in the elevation. By following this process with all the other points on the circle in the first projection, the necessary points are obtained through which to draw the ellipse in the elevation.

The shape of the joint line between the main pipe and the middle segment in the elevation may be obtained in a similar manner to that adopted for the determination of the ellipse. Thus, the points on the main circle in the plan are projected upwards into the elevation. Then, in the first projection, the distances are taken to left and right of the centreline $A''C''$ to each of the points on the joint line. These distances are then marked off above and below the centreline $A'C'$ on the corresponding lines in the elevation, thereby affording the necessary points through which to draw the shape of the joint line.

BRANCH PIPE OFF-CENTRE WITH MAIN AT COMPLEX ANGLE

The chief points to observe before attempting to solve the problem shown in Fig. 176 are
  (a) The branch pipe is connected to the main at an angle of 45° in the elevation and 45° in the plan.
  (b) The diameter of the branch is half of that of the main.
  (c) The branch enters at a position which is tangential to the main pipe.
This problem needs only one auxiliary projection for its solution. The projection and development of the pattern are shown in Fig. 177.

The projection is made at right angles to the centreline AB in the plan. It is important to note that the centreline of the branch in the plan meets the vertical centreline of the main circle at point A, and not at the centre O of the circle. This comment may seem obvious and unnecessary, but failure to appreciate this point is often the cause of error in the projection and also in the elevation when determining the shape of the joint line in that view.

To obtain the centreline of the branch in the projection, the two points A and B are projected forward, and a base or datum line drawn across and at right angles to the projection lines, as at $A''B''$. The height $A'C'$ is now taken from the elevation and marked off in the projection from A'' to C''. Then $B''C''$ represents the centreline of the branch, and in that position is lying flat in the plane of the paper.

Auxiliary and Double Projections

The two outside lines which represent the width or diameter of the pipe may now be drawn parallel to the centreline, and the end of the pipe cut square with the centreline through B''. A semicircle is drawn on the end of the pipe, divided into the usual six equal parts, and the points projected back to the end and onwards parallel to the centreline. The circular end of the pipe is then projected back to the

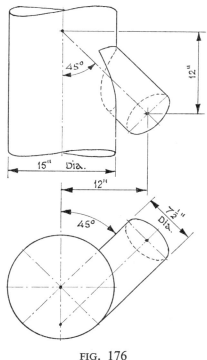

FIG. 176

plan to obtain the ellipse around point B. From the points on the ellipse, lines are drawn parallel to the centreline AB to meet the main circle.

Note that the outside line of the branch touches the circle at point D, which is on the centreline of the circle through O.

From the points thus obtained on the main circle, lines are drawn into the projection to meet the corresponding lines on the branch pipe in that view. Points are thereby afforded through which to draw the shape of the joint line. The pattern for the branch piece is now unrolled from the projection, as shown in Fig. 177. Also the contour of the hole in the main pipe is developed as shown to the left below the projection. The distance along the base line from D''' to E'' is made equal to the distance around the curve from D to E in the plan by taking the spacings therefrom and marking them off along the base line $D''E''$. Lines drawn from these points to meet the corresponding lines from the joint curve in the projection

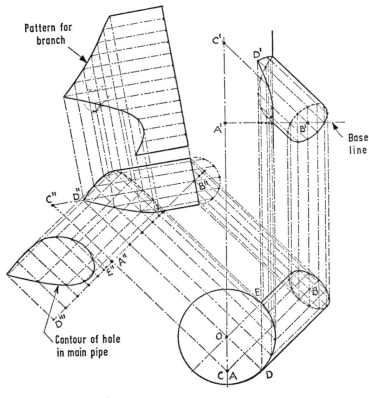

FIG. 177

give the necessary points through which to draw the shape of the hole.

It now remains to complete the ellipse and the joint curve in the elevation. First, the points around the ellipse in the plan are projected vertically upwards into the elevation. Then from the end of the pipe in the projection, the perpendicular distances from the base line $A''B''$ to each of the points on the end of the pipe are taken and marked off on the corresponding lines above and below the

Auxiliary and Double Projections

base line $A'B'$ in the elevation, thereby affording points through which to draw the ellipse.

Similarly, the points around the curve DE in the plan are projected vertically upwards into the elevation, and from the points on the base line $A''B''$ in the projection, the perpendicular distances to the points on the joint line are taken and marked off on the corresponding lines above the base line $A'B'$ in the elevation. Points are thereby afforded through which to draw the shape of the joint curve in the elevation.

TWIN BRANCH PIPES ON MAIN PIPE

The centrelines of the two branch pipes shown in Fig. 178 both intersect the centreline of the main pipe at A in the plan and at A' in the elevation. In this problem there are two joint lines to be considered: one between the branches and the main pipe, and the other between the two branches themselves. As the two branches

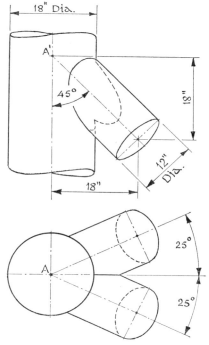

FIG. 178. *Pattern for one branch to be developed*

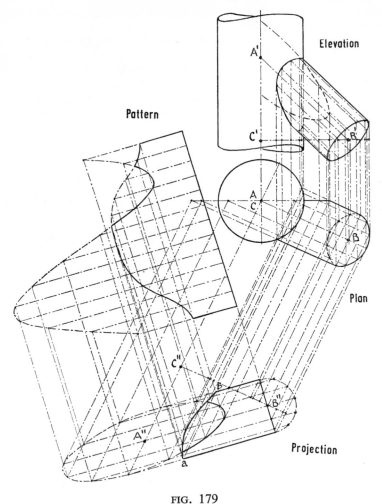

FIG. 179

are similar and symmetrically placed, one pattern will serve for both, and the solution will need only one auxiliary projection. The projection and pattern development are shown in Fig. 179.

Of the two branches, the lower one is selected, and from it the projection is made at right angles to its centreline AB. The two points A and B are projected downwards, and a base or datum line drawn across and at right angles to them in any convenient position, as at $B''C''$. Next, the height $C'A'$ is taken from the elevation and marked off from C'' to A'' in the projection. The line through A'' to B'' then represents the true length and true angle of the centreline of the branch pipe, and is lying flat in the plane of the paper. The diameter is now drawn parallel to the centreline of the pipe, and the end at B'' cut square with the centreline.

The next step in the process of solution is to determine the joint line in the projection. Since the joint line in this case is a complex one, consisting of a portion from each of the two joint lines, the two full joint lines are determined independently in order to show more clearly the method of approach to the solution.

First, the semicircle is drawn on the end of the pipe at B'', divided into six equal parts, and the points thereon projected along the pipe parallel to the centreline. The circular end of the pipe at B'' is next projected back to the plan to obtain the ellipse—a straightforward exercise in projection. From the points on the ellipse, lines are now drawn parallel to the centreline AB to cut the circular plan of the main pipe.

Assume for the moment that the second branch pipe does not exist; then the problem becomes one of simple intersection. Thus from the points where the parallel lines cut the circle of the main pipe in the plan, lines are drawn into the projection to meet the corresponding parallel lines on the branch in that view. Points are thereby obtained through which the joint curve ab is drawn. This curve represents the joint line between the branch and the main pipe.

Again, assume for the moment that the main pipe does not exist, and that the intersection lies between the two branch pipes only. In the plan view, this results in a straight horizontal cut along the centreline through A. The lines from the points on the ellipse are continued onward to meet this cutting line, which represents the joint line between the two branch pipes. The true shape or this joint line will, of course, be an ellipse, and is obtained in the projection by drawing lines from the points on the cutting line into the projection to meet the corresponding parallel lines on the branch pipe. Points are thereby afforded through which to draw the full ellipse in the projection. It will be seen that only a small portion of this ellipse is required when forming the joint line between the two branches and the main pipe.

The pattern for the branch pipe is unrolled from the projection. In this example the full contours of both joint lines are shown in the pattern, but the parts required for the branch piece are drawn in full outline, while the remaining parts of the contour are shown chain dotted.

It now remains to determine the shape of the joint line and the ellipse in the elevation. First, the points on the ellipse in the plan are projected vertically upwards into the elevation. Then from the base line $B''C''$ in the projection, the perpendicular distances are taken to each of the points on the end of the pipe, and marked off above and below the ground line $B'C'$ in the elevation. Points are thereby afforded through which to draw the ellipse.

Next, the points on the joint line in the plan are projected vertically upwards into the elevation, and then from the base line $B''C''$ in the projection, the perpendicular distances are taken to each of the points on the joint line. These distances are marked off on the corresponding lines above the ground line $B'C'$ in the elevation, thereby affording the necessary points through which to draw the shape of the joint line in that view.

The pattern for the second branch piece will be the same as that for the one described above, but will need to be rolled in the opposite direction.

A SIMPLE ELBOW AT COMPLEX ANGLES

The example shown in Fig. 180 represents a simple pipe elbow with the upper limb at 60° in the left-hand view and at 45° in the right-hand view. A simple elbow of this kind is sometimes presented on

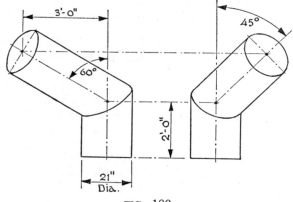

FIG. 180

a working drawing, among other details, and then the double angle as shown in the Figure raises the problem somewhat out of the elementary stage. Nevertheless, it requires only simple projection for its solution. The projections and development of the patterns are given in Fig. 181.

Auxiliary and Double Projections

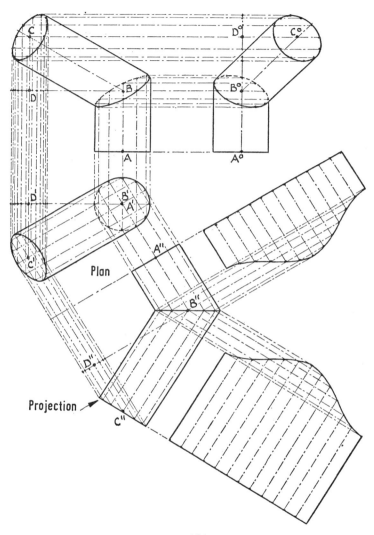

FIG. 181

The chief aim in making the projections is to obtain a view in which the centrelines of the pipes are both lying flat in the plane of the paper, when the view presents a simple elbow. First, the centreline AB is drawn and BC set off at 60° to the vertical. Then the centreline A^0B^0 is drawn and B^0C^0 set off at 45°. The next step is to drop a plan of the centrelines from the left-hand elevation. In this plan the centreline AB becomes a single point at $A'B'$, through which the horizontal ground line $B'D'$ is drawn. The point C is dropped into the plan, and the distance $D'C'$ made equal to D^0C^0 in the right-hand elevation. Then $B'C'$ is the centreline of the upper limb of the elbow in the plan.

In the plan it should be observed that the centreline $A'B'$ of the lower limb rises perpendicularly to the plane of the paper, and that the centreline $B'C'$ of the upper limb rises at an angle to the plane of the paper. From these conditions, a projection at right angles to the centreline $B'C'$ will present both centrelines lying flat in the plane of the paper, which is the view required for the development of the patterns.

Therefore points B' and C' are projected downwards at right angles to $B'C'$, and the distance $A''B''$ is marked off equal to AB from the left-hand elevation. A ground line $B''D''$ is drawn through B'' at right angles to $A''B''$, and the distance $D''C''$ marked off equal to DC from the left-hand elevation. Then the centrelines $A''B''$ and $B''C''$ are true lengths and are lying flat in the plane of the paper. The diameters of the pipe are drawn parallel to the centrelines, and the ends drawn or cut off square with the centrelines. The joint line then becomes a straight line through B''.

The patterns are now unrolled from the projection as shown in Fig. 181, and as they are examples of straightforward elementary parallel line development, the reader should experience no difficulty in following the procedure from the illustration. It is, however, more in keeping with the purpose of this chapter to determine the ellipses in the front and side elevations, the accuracy of which depends on the careful plotting of the points.

First, the ellipse in the plan is obtained by projecting the points on the end of the pipe at C'' in the projection back to the plan to meet the corresponding lines drawn from the circle at $A'B'$. Points are thereby afforded through which to draw the ellipse around C'.

Next, the points on the ellipse around C' are projected vertically upwards into the left-hand elevation. From the projection, the distances are taken from the ground line through D'' to the points on the edge at C'', and marked off on the corresponding lines above the ground line through D in the left-hand elevation. Points are thereby obtained through which the ellipse is drawn around the point C.

The ellipse around B is now obtained by projecting the points from the circle in the plan vertically upwards to the left-hand elevation. Then from the projection, the distances are taken from the edge at A'' to the joint line through B'', and marked off on the corresponding lines above the edge of the pipe at A in the left-hand elevation, thereby affording points through which the ellipse is drawn around point B. The outside lines are now drawn tangentially to the ellipses. These lines represent the diameter of the pipe, and should be spaced equal to it.

The ellipses in the right-hand elevation may now be dealt with. First, the points on the ellipse around C in the left-hand elevation are projected horizontally into the right-hand elevation. Then, from the plan, the distances are taken from the ground line through D' to the points on the ellipse around C', and marked off horizontally on the corresponding lines from the line through D^0. Points are thereby afforded through which the ellipse is drawn around C^0.

Finally, the points on the ellipse around point B in the left-hand elevation are projected horizontally into the right-hand elevation. From the plan, the distances above and below the ground line $D'B'$ to each of the points on the circle are taken and marked off on the corresponding lines to left and right of the vertical line A^0D^0 in the right-hand elevation. Points are thereby afforded through which the ellipse is drawn around B^0. The outside lines are now drawn tangentially to both ellipses to represent and be spaced equal to the diameter of the pipe. The diameter of the vertical portion of the pipe, A^0B^0, should also be drawn tangentially to the lower ellipse at B^0.

A Y-PIECE INCLINED AT THE JOINT

From some points of view the Y-piece shown in Fig. 182 is similar to the elbow shown in Fig. 180, except that two branch limbs are joined to the vertical pipe instead of one. The solution of the problem however, is not quite so simple as the previous example. In this case a plan is dropped from the right-hand elevation, and a projection made from the plan. The solution, with pattern developments, is shown in Fig. 183. In this figure the lobster-back bends on the two branches are ignored.

The vertical centreline AB in the right-hand elevation becomes a single point at A',B' in the plan, and is the centre of the circle which represents the plan of the vertical pipe AB. The two centrepoints

C and D are next dropped from the elevation into the plan to obtain the corresponding points C' and D'. The distances $E'C'$ and $E'D'$ are made equal to E^0C^0 and E^0D^0 from the left-hand elevation. Then $B'C'$ and $B'D'$ represent the centrelines of the two branches in the plan.

A projection of the vertical pipe and the lower branch is now made at right angles to the centreline $B'C'$. First, the centrepoint at A',B' is projected downwards at right angles to $B'C'$, and the distance $A''B''$ marked off equal to AB from the elevation. A ground

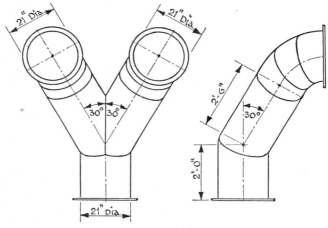

FIG. 182

line $B''F''$ is drawn through B'' at right angles to $A''B''$. A line from the centrepoint C' is now drawn into the projection, and the distance $F''C''$ marked off equal to FC from the elevation. The centrelines $A''B''$ and $B''C''$ now represent their true lengths and are lying flat in the plane of the paper. The pipe diameters are now drawn parallel to the centrelines, and the ends at A'' and C'' are drawn or cut across at right angles to the respective centrelines. The joint lines have now to be determined.

Assume for the moment that only the two limbs $A''B''$ and $B''C''$ exist. Then the joint between them would represent a straight line through B''. However, the presence of the other limb necessitates the plotting of two elliptical curves in the projection, one representing the joint line between the two branches and the other the joint line between the vertical pipe and the alternative branch. In the plan, the joint between the two branches is represented by the straight line $G'E'$ through the centre A',B'.

Auxiliary and Double Projections

The next step in the solution is to obtain the ellipse in the plan around point C' by the usual method of projecting from the edge C'' all the points thereon back to the plan and marking off the various widths taken from the semicircle on the edge at C''. Then, from the

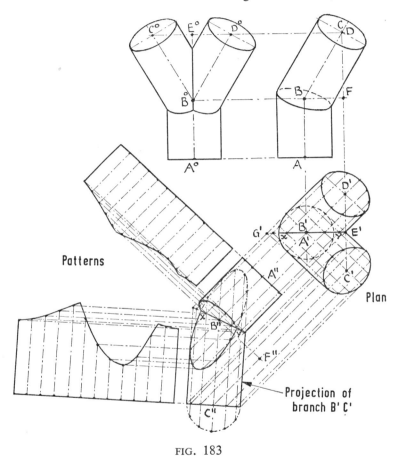

FIG. 183

points on the ellipse, lines are drawn parallel to the centreline $B'C'$ to cut the circle around the centre A',B'. This, in effect, divides the circle into twelve equal parts, as shown in the Figure. The divisions, however, need to be carefully checked as even a slight inaccuracy in drawing the lines can result in the divisions being anything but equal. The lines cut the straight joint line $G'E'$ in points which are needed in the next move to obtain the ellipse in the projection.

From the points on the semicircle at C'' in the projection, lines are now drawn parallel to the centreline $C''B''$ to cut through and beyond the straight joint line through B''. Next, from the points on the joint line $G'E'$ in the plan, lines are drawn into the projection parallel to the centreline $A''B''$ to meet the corresponding lines just drawn on the other limb parallel to $C''B''$. Points are thereby afforded through which the full ellipse is drawn around the centrepoint B'' as shown in Fig. 183. The full ellipse represents the joint between the two inclined branches only. Since the branches also intersect the vertical pipe, only half of the ellipse is required, as shown in full line in the diagram.

The joint line between the vertical pipe and the inclined limb $B'C'$ is represented in the plan by the corresponding half circle on xy. This half circle is represented in the projection by that portion of the straight joint line shown in full from point x through point y to the bottom extremity of the joint line.

The joint line between the vertical pipe and the other inclined limb on the centreline $B'D'$ is represented in the plan by the opposite half circle on xy. This half circle is represented in the projection by the semi-ellipse on xy, which is obtained by drawing lines at right angles to the centreline $B''A''$ from the points on the straight joint line from x through y. Where these lines meet or cross the corresponding parallel lines on the pipe $B''A''$, points are afforded through which the semi-ellipse is drawn.

The patterns for the vertical pipe and the inclined limb $B'C'$ are unrolled from the projection and it is assumed that, since the development is on straightforward parallel line principles, the reader at this stage should have no difficulty in following the procedure from the illustration. The pattern for the other limb, $B'D'$, is the same as that for the limb $B'C'$.

ANOTHER Y-PIECE WITH ONE LIMB AT DIVERSE ANGLE

The example of pipework shown in Fig. 184, composed of equal-diameter pipes, has diversion valves for regulating the direction of the air flow. This example is given chiefly for the exercise of determining the joint lines between the branches. When the joint lines are correctly obtained, the development of the patterns becomes a simple exercise in the application of the parallel line method.

In Fig. 185 the three limbs of the branch piece are shown without the conical connexion at the bottom and the lobster-back bends at the top. The bottom edge of the vertical pipe is cut horizontally in order to simplify the subsequent processes of locating the joint lines between the branches. In this Figure the full joint lines are

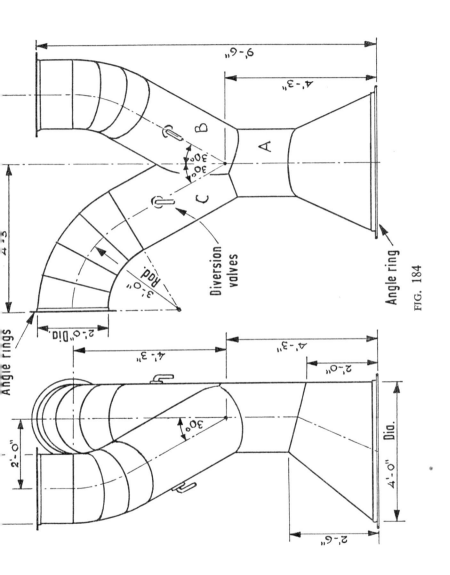

FIG. 184

shown, two as complete ellipses and one as a straight line. The portions of the joint lines which can be seen in the right-hand elevation in Fig. 184 are shown in full line in Fig. 185, with the remainder of the joint lines chain dotted.

To simplify the approach to the determination of the joint lines, each pair of pipes is considered separately, as in Fig. 186 between pipes *A* and *C*, Fig. 187 between pipes *A* and *B*, and Fig. 188

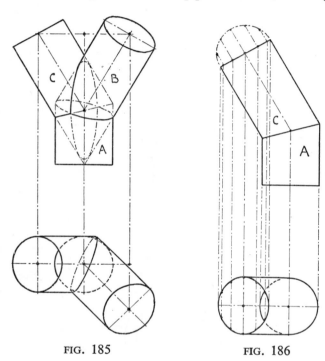

FIG. 185   FIG. 186

between pipes *B* and *C*. The combined operations are shown in one illustration in Fig. 189.

First, the two pipes *A* and *C* present a simple elbow as shown in Fig. 186, with the joint line represented by a straight line at the elbow. The plan of pipe *A* is a circle, and the top end of pipe *C* falls as an ellipse in the plan. The method of dropping the ellipse from the elevation into the plan should readily be followed from the illustration as the process is quite an elementary operation.

The second joint line between pipes *A* and *B*, shown in Fig. 187, presents an ellipse around the centrepoint *N* in the elevation. The

Auxiliary and Double Projections 223

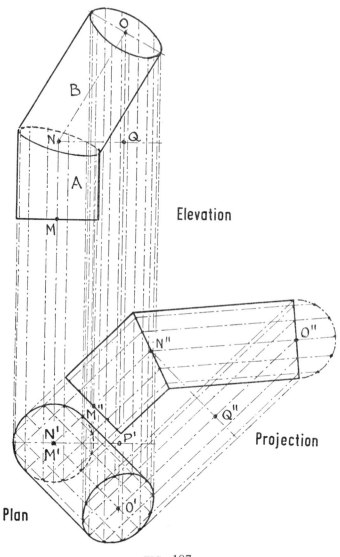

FIG. 187

first step towards obtaining this joint line is to drop a plan from the elevation, in which the distance $P'O'$ is made equal to $N'P'$, thereby making the angle $P'N'O'$ equal to 45°.

A projection is then made at right angles to the centreline $N'O'$ by projecting point N' forward and marking off the distance $M''N''$ equal to MN from the elevation. The point O' is now projected forward, and a ground line, $N''Q''$, drawn at right angles to the direction of projection. The distance $Q''O''$ is marked off equal to QO from the elevation. Then the centrelines $M''N''$ and $N''O''$ represent their true lengths and are lying flat in the plane of the paper. The diameters of the pipes are now drawn parallel to the centrelines, and the ends, at M'' and O'', are cut off square to the centrelines. The joint line then presents a straight line through N''.

Next, a semicircle is drawn on the end of the pipe at O'', divided into six equal parts, and the points thereon projected back along the pipe to the joint line. From the points on the joint line the lines are continued parallel to the centreline $M''N''$ back to the circle in the plan, which is thereby divided into twelve equal parts similar to those on the semicircle. The points on the circle must be checked for accuracy to ensure that the divisions are equal as shown in Fig. 187. Now, the circular edge at O'' in the projection is projected back to the plan to obtain the ellipse around point O'.

To obtain the ellipse around point N in the elevation, the points on the circle in the plan are projected upwards into the elevation. Then from the projection, the distances from the points on the bottom edge at M'' to the joint line at N'' are taken and marked upwards on the corresponding lines in the elevation from the bottom edge at M. Points are thereby afforded through which the ellipse is drawn around the point N. This ellipse represents the joint line between the two pipes A and B.

The ellipse around point O in the elevation may be obtained in a similar manner. The points around the ellipse at O' in the plan are first projected into the elevation, and then the distances in the projection from the ground line at Q'' to the top edge at O'' are taken and marked off on the corresponding lines in the elevation from the ground line through Q to the ellipse at the top. Points are thereby afforded through which the ellipse is drawn.

The third joint line between pipes B and C requires two projections in addition to the plan in order to determine its elliptical shape in the elevation (Fig. 188). The plan of these two pipes is drawn first, locating point O' in a similar manner to that shown in Fig. 187. Point R' is dropped vertically down from R in the elevation to the horizontal centreline $R'N'$.

The next step is to obtain the first projection from the plan. This

Auxiliary and Double Projections

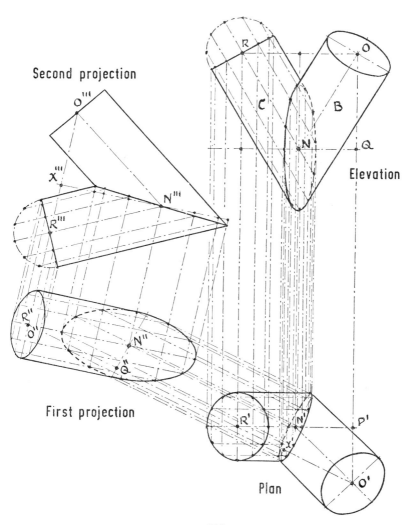

FIG. 188

is taken in the direction through the centrepoints O' and R'. Since these points lie at the same vertical height in the elevation, point R' will lie directly behind point O' in the projection, as at O'',R'', which is represented by a single point. The centrepoint N' is also projected forward parallel to $O'R'$, and a ground line drawn across and at right angles to the direction of projection as at N''. Then on the projection line from $O'R'$, the distance $Q''O''$ is marked off from the ground line and made equal to the height QO from the elevation. Also in the plan, from the centrepoint N' the datum line $N'X'$ is drawn at right angles to $O'R'$.

It is important now to observe that in the first projection the centreline $N''O''$ lies directly in front of the centreline $N''R''$, and in consequence the single line from N'' to the centrepoint O'',R'' represents both centrelines. Moreover, the two points O'' and R'', while being represented by the single point in the first projection, are really separated by a distance equal to the distance from O' to R' in the plan, the line joining them being at right angles to the plane of the paper.

A second projection now made at right angles to the double centreline $O''N''$–$R''N''$ results in the two centrelines $O'''N'''$ and $N'''R'''$ lying flat in the plane of the paper. Thus, the point N'' is projected forward to N''', and the datum line $N''''X''''$ drawn at right angles to the direction of projection. The centrepoint O'',R'' is next projected into the second projection, and the distances $X'''O'''$ and $X'''R'''$ made equal respectively to $X'O'$ and $X'R'$ from the plan. Then the centrelines $N'''O'''$ and $N'''R'''$ represent their true lengths and show the true angle between them in the second projection.

The diameters are now drawn parallel to the centrelines, and the ends of the pipes at O''' and R''' are cut off square or at right angles to the centrelines. The joint line through N''' now bisects the true angle of the centrelines. It should be noted that the joint line does not lie on the datum line $X'''N'''$.

The series of ellipses representing the joint line in the first projection, the plan and the elevation, may now be obtained by projecting back from the second projection. A semicircle is first drawn on the end of the pipe at R''', divided into six equal parts, and the points projected along the pipe parallel to the centreline to meet the joint line through N'''. The points on the joint line are then projected back to the first projection.

Next, the circular end of the pipe at R''' is projected back to the first projection to obtain the ellipse around the centrepoint O'',R''. From the points on this ellipse, lines are drawn parallel to the centreline $R''N''$ to meet the lines drawn from the joint line through N'''. Points are thereby afforded through which the ellipse around the

Auxiliary and Double Projections

centrepoint N'' is drawn. This ellipse represents the joint line in the first projection.

To determine the ellipse representing the joint line in the plan, the points around the joint ellipse in the first projection are projected back to the plan as far as the datum line through $N'X'$. The next move requires a fair degree of care to ensure the accuracy of the ellipse around point N'. From the second projection, the short distances from the datum line $N'''X'''$ to the joint line through N''', above and below, are taken and marked off on the corresponding lines above and below the datum line $N'X'$ in the plan. Points are thereby afforded through which the ellipse is drawn around the centrepoint N'.

There are two methods whereby the elliptical joint line in the elevation may now be obtained, but first, common to both, the points on the joint ellipse in the plan are projected into the elevation. Then, by the first method and from the first projection, the distances are taken from the datum line $Q''N''$ to the points on the elliptical joint line, above and below, and marked off on the corresponding lines above and below the ground line or datum line QN in the elevation. Points are thereby obtained through which the elliptical joint line is drawn.

For the second method, the circular edge at R in the elevation is dropped into the plan to obtain the ellipse around the centrepoint R'. From the points on the ellipse, lines are drawn parallel to the centreline $R'N'$ to cut the joint ellipse around N'. From the points where these lines cut the joint ellipse, vertical lines are drawn into the elevation to meet a corresponding set of lines from the top edge at R. Where these two sets of lines meet, points are afforded through which the joint ellipse is drawn in the elevation.

There is an important point of difference in these two methods which should be noted. In the first method the points on the joint ellipse in the plan were obtained from the first and second projections, and these points were projected into the elevation. The ellipse in the elevation was then located on these lines. In the second method, the ellipse around the centrepoint R' was obtained from the elevation, and parallel lines drawn from the points on the ellipse in the plan to cut the joint ellipse around N'. Now, the points where these parallel lines cut the joint ellipse do *not* necessarily occur at the same points as those by which the joint ellipse was obtained from the first projection. Nevertheless, in the example illustrated in Fig. 188 the points occur so close together that any difference would be difficult to discern.

As already stated, the three joint lines dealt with separately in Figs. 186, 187 and 188 represent independent solutions of the determination of the joint lines between the three pipes shown in Figs. 184 and 185. The combined solution of the problem is given in Fig. 189, which represents the simple combination of the three previous Figures.

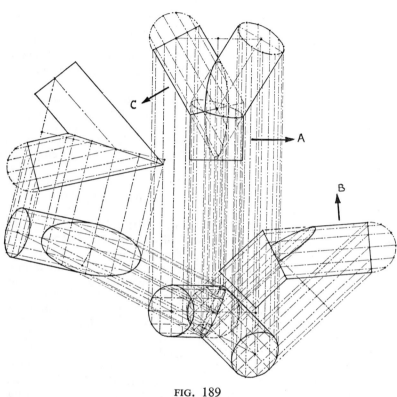

FIG. 189

The patterns for the three pipes may be unrolled in the directions indicated by the arrows in Fig. 189. Once the joint lines have been determined the patterns may be obtained by straightforward parallel line development and it is assumed that this would present no difficulty to the reader at this stage. Therefore in order to avoid further congestion of the lines the development of the patterns has not been shown in the Figure.

FEED CHUTE TO ROTARY SIEVING MACHINE

The chute illustrated in Fig. 190 represents a feed inlet connexion to a rotary sieving machine. The circular disc is a fixture which carries the bearing for the centre spindle and also contains the feed hole around which the chute flange is fixed. The two sides of the

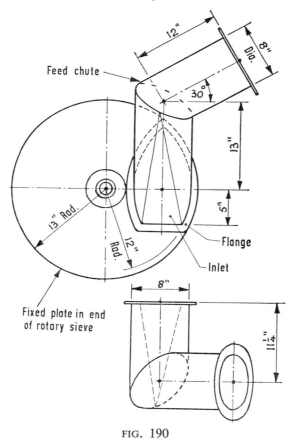

FIG. 190

flange conform to the curve of the fixed plate while the bottom is flattened to facilitate the flow of material into the sieve. The bottom part of the feed chute forms an elbow with the upper cylindrical part which receives the material from a previous operation.

Two auxiliary projections are required for the complete solution of the problem. First, a view in the direction of the arrow in the elevation is taken, Fig. 191, looking down the axis of the cylindrical pipe. This pipe then becomes a circle in the first projection, and the axis $B'C'$ becomes the centrepoint B'',C'' of the circle. Next, the distance $B''X''$ is marked back along the projection line and made

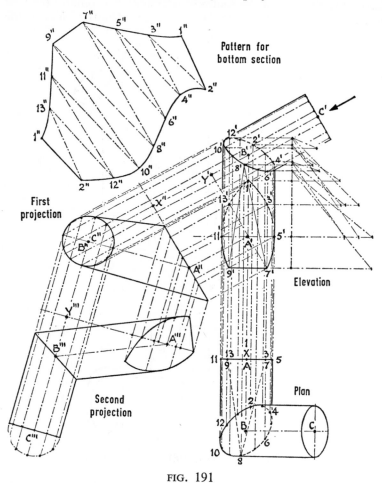

FIG. 191

equal to BX from the plan. The ground line $A''X''$ is then drawn at right angles to $B''X''$, and the point A' projected from the elevation to obtain point A'' in the first projection. The line joining A'' and B'' now represents the centreline of the bottom section of the chute. To complete the first projection, points $1'$ and $7'$ from the elevation

Auxiliary and Double Projections 231

are projected to the ground line $A''X''$, and from the ground line are drawn tangentially to the circle centre B'',C''.

The second projection is a view obtained from the first projection at right angles to the centreline $A''B''$. Thus, the two points A'' and B'' are projected downwards at right angles to $A''B''$, and a ground line $A'''Y'''$ drawn across and at right angles to the direction of projection. The distance from Y' to B' is now taken from the elevation and marked off from Y''' to B''' in the second projection. Also, the distance from B' to C' is taken from the elevation and marked off from B''' to C''' in the second projection. The two centrelines $A'''B'''$ and $B'''C'''$ now present their true lengths and are lying flat in the plane of the paper.

To complete the second projection, the pipe diameter is drawn parallel to the centreline $B'''C'''$, and the end of the pipe at C''' is drawn or cut square with the centreline $B'''C'''$. The joint line through B''' bisects the angle $A'''B'''C'''$. To obtain the shape of the inlet end of the chute around the centrepoint A''', this shape in the elevation is first divided into a convenient number of parts as from $1'$ to $13'$, the points projected to the ground line $A''X''$ in the first projection, and from there into the second projection. Next, in the elevation, the perpendicular distances from the ground line $A'Y'$ are taken to the points on the inlet, above and below, and marked off on the corresponding lines above and below the ground line $A'''Y'''$ in the second projection. Points are thereby afforded through which the shape of the inlet is drawn. The outside lines of the inlet section are now drawn from the elbow to the inlet shape.

In order to develop the pattern for the lower or inlet section of the chute, it is necessary to apply the method of triangulation, and hence a plan and elevation are required from which the plan lengths may be triangulated against vertical heights. For this purpose any two adjacent views in Fig. 191 may be used, provided one of them is used as the plan and the other as the elevation. However, in this example use is made of the normal plan and elevation, and this now requires the completion of the joint ellipse in both views.

The joint ellipse is obtained by projecting back, or in the reverse order from the second projection. Thus a semicircle is described on the end of the pipe at C''' and divided into the usual six equal parts. These points are projected back along the pipe, through the joint line at B''', and on to the circle around B'',C'' to divide that circle into twelve equal parts. The equal parts should be similar to those on the semicircle and should be checked for equality. Then from the points on the circle in the first projection, lines are drawn back to the elevation as far as the end of the pipe at C'. Next, from the second projection, distances are taken from the end of the pipe at

C''' to the joint line at B''', and marked back on the corresponding lines from the end of the pipe at C'. Thereby points are afforded through which the joint ellipse is drawn around the centrepoint B'.

To obtain the joint ellipse around the centrepoint B in the plan, the points on the joint ellipse in the elevation are dropped vertically downwards. Then from the first projection, the distances from the ground line at X'' are taken to the points on the circle and marked off downwards on the corresponding lines from the ground line at X in the plan. Points are thereby obtained through which the ellipse is drawn around the centrepoint B in the plan.

The ellipse around the centrepoint C in the plan may readily be obtained by dropping the circular edge at C' from the elevation into the plan by the usual method.

The pattern for the bottom section of the chute is now obtained by the method of triangulation. The plan and elevation are divided into a number of convenient parts as shown in Fig. 191. Detailed directions for the development are not given, as it is assumed that the reader at this stage is sufficiently familiar with the method of triangulation to experience no difficulty in following the development from the illustration. There are two important points, however, which should be observed in the development. First, the true spacings for the inlet end of the pattern should be taken direct from the elevation, as from 1' to 3', 3' to 5', and so on. Second, the true spacings for the joint line at the opposite end cannot be obtained direct from the plan or the elevation, but each spacing must be triangulated by placing its plan distance at right angles to its vertical height. The diagonal will then give the true spacing.

COMPLEX CONNEXION FROM RECTANGLE TO CIRCLE

The problem contained in the example shown in Fig. 192 is to connect the rectangular hole, 21 in. long by 6 in. wide, to the circular hole of 12 in. diameter which lies just above the rectangular hole and is almost in the same vertical plane. For the transition from the rectangle to the circle it is clear that a transforming piece must be included in the construction. Although the design and construction may be varied in many ways, the arrangement shown in Fig. 192 is probably the most efficient. The development of the pattern for the transforming piece is shown in Fig. 193.

The front elevation of the transforming piece only is shown in Fig. 193, and an auxiliary projection is made in the direction of the rectangular base as indicated by the arrow. The making of the auxiliary projection is a fairly straightforward process. The main points to be observed are that the length of the centreline BC of the

Auxiliary and Double Projections

cylindrical portion is made equal to that given in the side elevation, that the projected distance from the edge of the cylindrical portion to the centre *A* of the base is made equal to that of the side elevation, and that the joint line through B bisects the angle made by the centre-lines *AB* and *BC*.

For the purpose of developing the pattern, the front elevation is used as the plan and the projection as the elevation. Thus, viewing

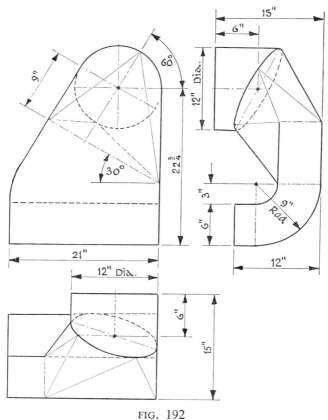

FIG. 192

the new plan from the direction of the arrow, the projection falls readily into line as the new elevation.

The circle in the plan is divided into twelve equal parts as shown in the Figure, and the points thereon are projected to the joint line at *B* in the elevation. From these points lines are drawn at right angles to the joint line, and the corresponding semi-widths of the circle are marked off on these lines to obtain the semi-ellipse 1′6′10′. The divisions between the points on the semi-ellipse are the true spacings required for the joint edge in the pattern.

With the new plan and elevation divided up for triangulation and a vertical height line erected in the direction of the projection as

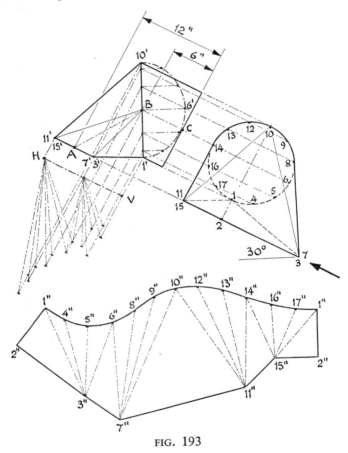

FIG. 193

shown in the Figure, the process of developing the pattern should offer little or no difficulty to the reader at this stage.

A SIMPLE OFF-SET PIPE

The off-set pipe shown in Fig. 194 is perhaps an elementary example compared with most of those already dealt with, but is included

Auxiliary and Double Projections 235

mainly to serve as an introduction to the two problems which follow. The auxiliary projection necessary for its solution, and the pattern development are shown in Fig. 195.

The auxiliary projection is made from the plan at right angles to the centreline $A'B'$ by projecting points A' and B' forward into the projection and drawing a ground line in any convenient position through A'' at right angles to the direction of projection. Next, the

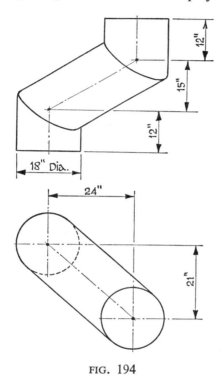

FIG. 194

distance CB is taken from the elevation and marked off from C'' to B'' in the projection. The centrelines $D''A''$, $A''B''$ and $B''E''$ represent their true lengths and are lying flat in the plane of the paper. The pattern for the middle section is now unrolled at right angles to the centreline $A''B''$ as shown in Fig. 195.

The two elliptical joint lines in the elevation may be obtained by projecting the points from the two circles in the plan into the elevation, and then taking the distances from the two edges D'' and E'' in the projection to the respective joint lines, and marking them off on the corresponding lines in the elevation from the two edges *D* and *E*. Points are thereby obtained through which the two ellipses around the centrepoints *A* and *B* are drawn.

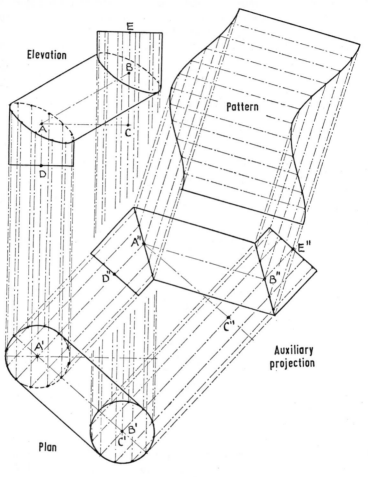

FIG. 195

ANOTHER OFF-SET PIPE

The example shown in Fig. 196 represents an off-set pipe somewhat similar to that of Fig. 194, but with one important difference. In the previous example the off-set middle section connects two vertical

Auxiliary and Double Projections

pipes, while the middle section in Fig. 196 connects one vertical pipe to a horizontal pipe in an off-centre position. In this case a simple auxiliary projection is not sufficient for the full solution of the problem. The necessary projections and the development of the pattern for the middle section are shown in Fig. 197.

As in most problems of this kind the solution depends chiefly on the correct determination of joint ellipses before the patterns can

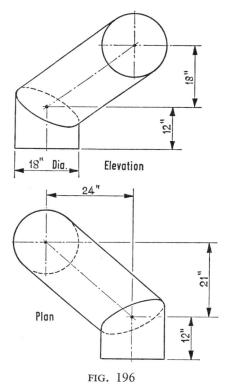

FIG. 196

be unfolded. For example, the ellipse around the centre B' in the plan must be determined before the corresponding ellipse around the centrepoint B'' in the first projection can be obtained, and in turn before the pattern for the middle section can be unrolled.

The first step towards the determination of the ellipse around B' in the plan is to make an auxiliary projection of that elbow from the elevation. Thus the two points A and B in the elevation are projected at right angles to the centreline AB, and a ground line A^0F^0 is drawn parallel to AB in any convenient position. Then, from the plan, the

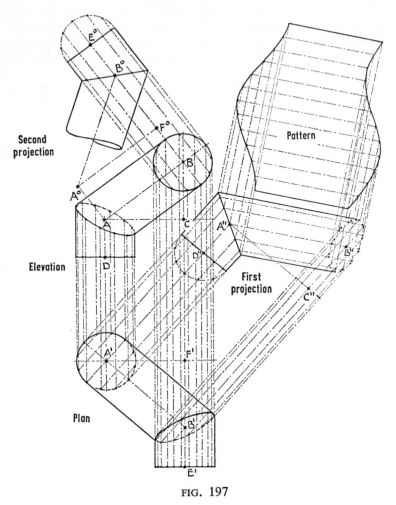

FIG. 197

distance $F'B'$ is taken and marked off from F^0 to B^0 in the second projection. Also from the plan, the distance $B'E'$ is taken and marked off from B^0 to E^0 in the second projection. The centrelines A^0B^0 and B^0E^0 now represent their true lengths and are lying flat in the plane of the paper. The pipe diameters are now drawn parallel to their respective centrelines and the top edge of the pipe cut square or at right angles to the centreline B^0E^0. The joint line through B^0 bisects the angle $A^0B^0E^0$.

Auxiliary and Double Projections

In the next stage of the process a semicircle is described on the top edge of the pipe at E^0 and divided into six equal parts. From the points on the semicircle, lines are drawn parallel to the centreline E^0B^0 to cut through the joint line at B^0 and onwards to the circle around the centrepoint B to divide the circle into twelve equal parts similar to those on the semicirle at E^0. From the points on the circle around B, lines are now dropped into the plan to meet the bottom edge of the pipe at E'. Next, from the second projection, the distances from the top edge at E^0 to the joint line at B^0 are taken and marked back on the corresponding lines from the bottom edge at E'. Points are thereby afforded through which the ellipse around B' is drawn.

Note. The term "corresponding lines" refers to lines which follow in sequence from one view to another through the various projections, including the plan and elevation.

From the points on the ellipse at B', lines are now drawn into the first projection. Then from the elevation, heights are taken from the ground line through C to the points around the circle at B and marked off along the corresponding projection lines from the ground line $A''C''$ in the first projection. Points are thereby obtained through which the ellipse around the centrepoint B'' is drawn. The centrepoint B'' is also obtained by taking the height from C to B in the elevation and marking it off from C'' to B'' in the first projection. Similarly, the distance $A''D''$ is obtained by marking it off equal to AD from the elevation.

From the illustration it will be seen that the lower limb presents an ordinary elbow in the first projection with a straight joint line through A''. The pattern for the middle section is now unrolled from the first projection by straightforward parallel line development. One point, however, must be observed in the development. The parallel lines drawn from the joint line at A'' do not necessarily meet the points already established on the ellipse around B'', and for the pattern development it is important to use the points where the parallel lines meet the curve of the ellipse and ignore the points by which the ellipse was determined.

To complete this problem, the ellipse around the centrepoint A in the elevation may now be obtained by taking the heights from the edge D'' in the first projection to the joint line at A'', and marking them off on the corresponding lines from the bottom edge at D in the elevation. Points are thereby obtained through which the ellipse around A is drawn.

An Alternative Solution Fig. 198 represents an alternative solution of the off-set pipe problem shown in Fig. 196. In this case there is only one projection, though the solution as a whole may appear to be somewhat more complex than that in Fig. 197.

The first stage is to set out the plan and elevation to the particulars given in Fig. 196, though the ellipses cannot yet be determined. Next, the centre lines in the projection are set out as in Fig. 197, but with the addition of the part $B''E''$ which is parallel to $A''C''$. The elbow at A'' is now drawn, and the semicircle at D'' described. This semicircle is divided into six equal parts and lines drawn through the points thereon to the joint line at A'' and also back to the circle at A'. The circle is thereby divided into similar divisions to those on the semicircle.

From the points on the circle at A', lines are drawn into the elevation. The ellipse around the centre A is then determined as in the previous example by taking the heights from the base line at D'' to the joint line at A'' in the projection, and marking them off on the corresponding lines from the base line at D in the elevation. This process locates sufficient points through which the ellipse can be drawn.

In the next stage the procedure deviates from the solution shown in Fig. 197. Lines are drawn parallel to the centreline AB from the points on the ellipse at A to cut the circle at B. From the points thus obtained on the circle, lines are dropped into the plan to cut a corresponding series of lines drawn parallel to the centreline $A'B'$ from the points on the circle at A'. Points are thereby afforded through which the ellipse is drawn around the centre B'.

From the points on the ellipse at B', lines are drawn then into the projection to meet another series of lines drawn parallel to the centreline $A''B''$ from the points on the joint line at A''. The ellipse around point B'' may now be drawn through the points where the corresponding lines meet.

To complete the projection, lines are drawn parallel to the centreline $B''E''$ from the points on the ellipse around B'', to meet a series of lines drawn from the pipe edge at E' in the plan. The ellipse around the centrepoint E'' in the projection may now be drawn through the points where the corresponding lines meet. The pattern for the middle section is unrolled from the projection at right angles to the centreline $A''B''$ as shown in the Figure.

Success in plotting well-formed ellipses depends very largely on the care and accuracy with which the construction lines are drawn. Any slight deviation from the parallel, or inaccuracy in starting lines from their exact points of origin, will result in considerable inaccuracy in the shapes of the ellipses.

Auxiliary and Double Projections

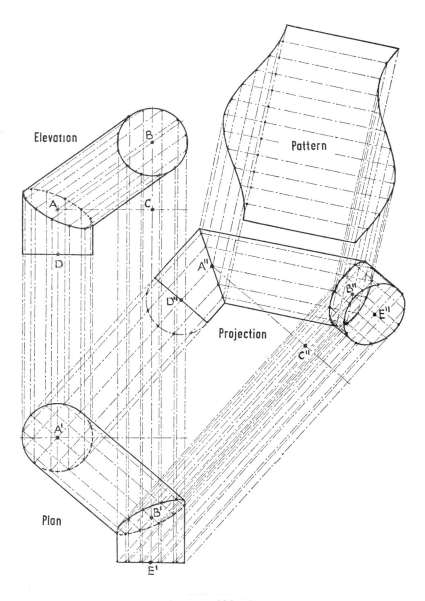

FIG. 198

A PIPE ENCIRCLING A SQUARE COLUMN

The pipe encircling a square column shown in Fig. 199 presents a somewhat similar problem to that shown in Fig. 196, inasmuch as each segment of the pipe meets the adjoining segments at different angles. In Fig. 199 all the segments are the same, so that one pattern will serve for all. In the illustration, one revolution of the pipe is

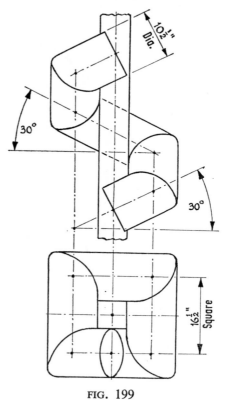

FIG. 199

shown, containing three full segments and two halves. The solution of the problem, including the pattern development, is shown in Fig. 200.

The first projection is made along the centreline of the middle segment in the direction from C to B. This segment then becomes a circle in the projection as seen around the centre B', C'. The distance $B'E'$ is next marked off along the projection line, and made equal to B^0E^0 from the plan. The line $A'D'$ is drawn through E'

Auxiliary and Double Projections

at right angles to $B'E'$. The two centrepoints A and D are now projected into the first projection to obtain the two points A' and D'. The second projection is then made from the first projection at right angles to the centreline $C'D'$. The two points C' and D'

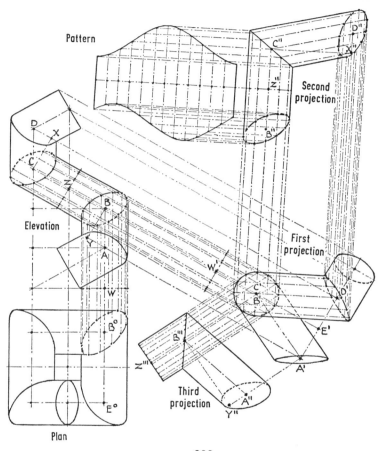

FIG. 200

are projected forward and the distance $B''C''$ marked off in any convenient position along the line from point C', and made equal to BC from the elevation. The line $C''X''$ is now drawn at right angles to $B''C''$. Next, in the elevation, the line CX is drawn at right angles to BC. The height XD is taken, and from X'' in the second projection the height $X''D''$ is marked off. The two centrelines

$B''C''$ and $C''D''$ now represent the true lengths of the centrelines BC and CD from the elevation and are lying flat in the plane of the paper. The pipe diameters are now drawn parallel to the centrelines and the joint line through C'' is drawn as a straight line as shown in the Figure.

Before the pattern can be unrolled from the segment $B''C''$, the ellipse around B'' must be determined. The straight joint line at C'' represents the joint between the segments $B''C''$ and $C''D''$ which are now lying flat in the plane of the paper, but the next segment which joins the pipe at B'' would not be lying flat, but would be rising at an angle to the plane of the paper, thereby producing an ellipse in the second projection at B''.

The next stage in the solution is to divide the circle in the first projection into twelve equal parts in the positions shown in the Figure. Lines are then drawn from the points on the circle into the second projection to meet the joint line at C''. Also, a line is drawn at right angles to the centreline $B''C''$ through a point Z'' which is midway between B'' and C''.

A third projection is now necessary in order to obtain the ellipse at B'' in the second projection. It should again be noted that the centrelines $B''C''$ and $C''D''$ in the second projection represent the true projection of $B'C'$ and $C'D'$ from the first projection, and that the joint line at C'' is in consequence represented as a straight line. A similar true projection is needed for the lower elbow between $C'B'$ and $B'A$. This projection is therefore made at right angles to the centreline $B'A'$.

Thus, points B' and A' are projected downwards at right angles to $B'A'$ and the distance $B'''Z'''$ is marked off equal to $B''Z''$ from the second projection. The line $B'''Y'''$ is now drawn at right angles to $B'''Z'''$. Next, in the elevation, the line BY is drawn at right angles to BC. The distance YA is now taken, and from Y'' in the third projection the distance $Y''A''$ is marked off. The two lines $Z'''B'''$ and $B'''A'''$ represent the true lengths of the centrelines AB and BC from the elevation and are lying flat in the plane of the paper. The pipe diameters are now drawn parallel to the centrelines, and the joint line through B''' is drawn as a straight line. Also, the diameter is drawn across the pipe through Z'''.

The elbow at B''' in the third projection now serves to obtain the ellipse around B'' in the second projection. First, from the points on the circle in the first projection, lines are drawn into the third projection to cut the joint line at B'' and on to the diameter through Z'''. Then the distances from the diameter at Z''' to the joint line at B''' are taken and marked downwards on the corresponding lines from the diameter at Z'' in the second projection, thereby obtaining

Auxiliary and Double Projections

the points through which the ellipse is drawn around the centrepoint B''.

The pattern for the segment $B''C''$ is now unrolled from the second projection by straightforward parallel line development, as shown in the Figure, with the spacings along the girth line projected from Z''.

The ellipses representing the joint lines in the elevation may now be determined. The line through point Z, midway between points B and C, is drawn across the pipe at right angles to BC. From the points around the circle B',C' in the first projection, lines are drawn into the elevation to cut through the line Z. Now, from the second projection, the distances from the line at Z'' to the joint line through C'' are taken and marked off on the corresponding lines from the line at Z in the elevation to obtain the points through which the ellipse around the centrepoint C is drawn.

The ellipse around the centrepoint B is obtained in a similar manner by using the elbow in the third projection. The distances from the line at Z''', to the joint line through B''' are taken and marked off on the corresponding lines, this time downwards from the line at Z in the elevation, to obtain the points through which the ellipse around the centrepoint B is drawn.

The ellipse around B^0 in the plan may now be obtained from the elevation. First, the distance from B^0 to W is taken and marked off in the first projection from the centre B' to W'. The line through W' is drawn across and at right angles to the projection lines.

Note. The reason for taking the distance from B^0 to W in the plan is that point W lies on a convenient ground line. Any line could be taken at any distance from B^0 to serve as a ground line. The important point in this connexion is that in the first projection the distance $B'W'$ must be made equal to the distance B^0W from the plan.

The next step is to drop all the points on the ellipse around B from the elevation into the plan. Then, from the first projection, the distances from the cross line W' to the points on the circle are taken and marked off on the corresponding lines from the ground line at W to obtain the points through which the ellipse around the point B^0 is drawn.

In the plan the four corner ellipses are all the same. Therefore the curves of the ellipse at B^0 are repeated at the other corners as shown in the Figure.

To complete the remainder of the problem, the ellipse around the centrepoint D'' in the second projection is similar to the ellipse around the opposite point at B'', and may be obtained by drawing the parallel lines from the joint line at C'', and then transferring the distances along the corresponding lines from the vertical pipe $B''C''$ to the pipe $C''D''$. The points are thereby obtained through which the ellipse is drawn around point D''.

To obtain the ellipse at D' in the first projection, the points on the ellipse around D'' in the second projection are dropped parallel to the projection lines at B',C'' into the first projection to meet corresponding lines drawn from the points around the circle as shown in the Figure. The ellipse obtained at D' is fairly narrow, as will be seen in the diagram, but is, nevertheless, a true ellipse which represents the joint line in that view. The ellipse at A' on the opposite limb in the first projection is similar to the ellipse at D'.

AN OVERFLOW CHUTE

The feed chute and overflow chute shown in Fig. 201 represents a useful application of the problem of cross pipes which intersect to a depth of half their diameters. The feed chute, in the right-hand elevation, slopes downwards at an angle of 45°, while the overflow crosses at an angle of 30° to the horizontal. The material normally runs down the underside of the feed chute, but in the event of a choke or

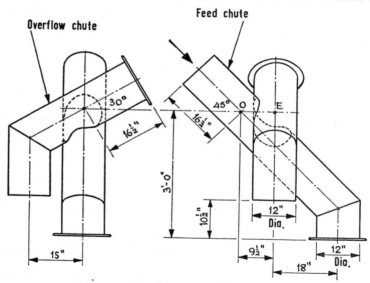

Overflow chute to intersect half way to centreline of feed chute

FIG. 201

Auxiliary and Double Projections

obstruction occurring at the bottom end of the chute, the material builds back or fills up the feed chute until it reaches the cross pipe, when it diverts down the overflow chute to be received in a spare bin or any other receptacle. The setting out of the problem and the pattern developments are shown in Fig. 202.

In setting out the elevations it should be noted that the two centrelines do not cross each other in the same position in both

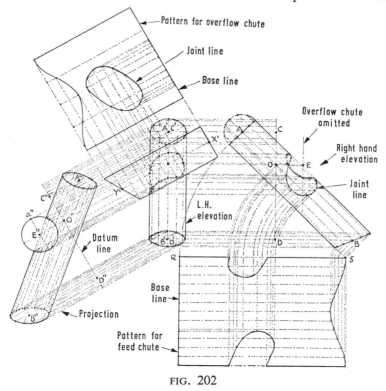

FIG. 202

views. In the right-hand elevation the centreline of the overflow chute is displaced to a distance OE from the corresponding point in the left-hand elevation at which the two centrelines cross.

To determine the distance OE an auxiliary projection is needed in the direction of the axis of the overflow chute. Thus, in Fig. 202 the projection is made in the direction $X'Y'$ from the left-hand elevation. A datum line $C''D''$ is drawn at right angles to the projection lines. This datum line corresponds to the vertical line CD, drawn through the centrepoint O in the right-hand elevation.

Next, the distance CA is taken from the right-hand elevation and marked off in the projection from point C'' to A'' along the line projected from A' in the left-hand elevation. Also, the distance DB is taken from the right-hand elevation and marked off in the projection from point D'' to B'' along the line projected from B' in the left-hand elevation as shown in the Figure. The line $A''B''$, passing through O'', then represents the centreline of the feed pipe in the projection, though it is not lying flat in the plane of the paper in that view.

The overflow pipe in the projection presents a circle, and in order that it shall intersect the feed pipe to half its depth, the centre of the circle must be moved along the projection line from O'' to E''. Then the distance $O''E''$ is taken and marked off in the right-hand elevation from O to E, which thereby determines the position of the overflow chute in that elevation.

The next step is to determine the ellipses around points A'' and B'' in the projection. To do this, a semicircle is drawn on the end of the feed chute at A, divided into the usual six equal parts, and lines drawn from the points on the semicircle back to the end of the chute at A. From the points on the end of the chute lines are then drawn into the left-hand elevation, and the ellipse plotted around the centrepoint A',C'. The plotting of this ellipse is a straightforward exercise, and at this stage should not need detailed instructions.

From the ellipse at A',C', the second ellipse around the centrepoint A'' is obtained by projecting the points from the ellipse at $A'C'$ into the projection. It is important to observe carefully the procedure in the next step. In the right-hand elevation, the points on the top end of the chute at A are projected back to the vertical datum line at C. Then the distances from the vertical line at C to the points on the end of the chute at A are taken and marked back on the corresponding lines from the datum line $C''D''$ in the projection, thereby determining points on the ellipse around A'' in the projection.

Next, from the points on the ellipse at A'', lines are drawn parallel to the centreline $A''B''$. In the left-hand elevation, lines are drawn parallel to the centreline $A'B'$ from the points on the ellipse at A'. Similarly, in the right-hand elevation, lines are drawn parallel to the centreline AB from the points on the end of the chute at A.

The next stage in the development is to determine the ellipses around the centrepoints B' and B''. This is done by projecting the points on the bottom edge of the chute at B into the left-hand elevation to meet the corresponding parallel lines in that view, thereby obtaining points on the ellipse at B'. From these points, lines are drawn into the projection to meet the corresponding

Auxiliary and Double Projections

parallel lines in that view, thereby obtaining points on the ellipse at B''.

The contours of the joint lines as seen in the left-hand and right-hand elevations are next determined by projecting the points on the semicircle at E'' back to the left-hand elevation, to meet the corresponding parallel lines in that view. Sufficient points are thereby obtained through which to draw the kidney-shaped joint line in the left-hand elevation.

To obtain the contour of the joint line in the right-hand elevation, the points on the joint line in the left-hand elevation are now projected into the right-hand elevation to meet the corresponding parallel lines in that view. Again, sufficient points are thereby obtained through which to draw the joint line in the right-hand elevation.

It remains now to develop the patterns for both chutes. The seam, in both cases, is arranged to occur on the top of the chute in order to provide a smooth-running surface on the bottom and facilitate the flow of the material.

The overflow chute is unrolled from the left-hand elevation as shown in Fig. 202. The divisions marked along the base line are not the usual equal spacings, but are taken from the circle in the projection. Beginning at point a'', the spacings are taken in clockwise rotation to the points where the parallel lines on the feed chute cross the circle, and marked off along the base line in the pattern. Then from the points thus marked off, lines are drawn at right angles to the base line. Next, from the points on the kidney-shaped joint line in the left-hand elevation, lines are drawn into the pattern to meet the corresponding lines drawn from the base line. Points are thereby afforded through which to draw the shape of the joint line in the pattern.

The curve in the pattern which represents the lower end of the overflow chute conforms to the usual shape for the angular cut across the pipe. In the development in Fig. 202 only four divisions have been used to locate the curve, in order to avoid congestion of lines with the plotting of the oval joint line.

The pattern for the feed chute may be unrolled from the right-hand elevation at right angles to its centreline AB. In Fig. 202, the development lines are swung through an angle of 45° to bring the pattern into a vertical position and keep it separate from the other views. Thus all the points on the joint line and the bottom edge are first projected at right angles to AB to the side of the chute. Then all the points on the side of the chute are swung round to the position RS, using S as the centrepoint. From the line RS the points are dropped vertically into the pattern.

Next, one of the spacings is taken from the semicircle on the top edge at *A*, and the usual twelve spaces are marked off downwards along the line from point *R*. Horizontal lines are now drawn from the base line at *R* to cross the vertical lines from *RS*. Where the horizontal lines cross the corresponding vertical lines obtained from the joint line in the elevation, points are afforded through which the contour of the joint line may be drawn in the pattern as shown in Fig. 202. It will be noted that as the longitudinal seam in the pipe is to be at the top, the contour of the joint line in this case is cut into two parts, one part occurring on each side of the pattern. The contour of the bottom edge in the pattern represents straightforward development and should readily be followed from the Figure.

A CYLINDRICAL CONNEXION TO AN OBLIQUE CONE

The example shown in Fig. 203 represents a cylindrical pipe connexion to an oblique conical reducing piece. At the top, the side of the cone is horizontal. In the left-hand elevation the branch pipe

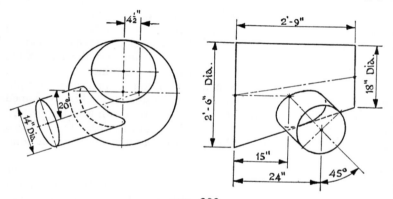

FIG. 203

connects to the cone at an angle of 20° to the horizontal, and in the right-hand elevation at an angle of 45° to the horizontal. This combination represents a typical unit in a system of ductwork.

The solution of the problem, which requires two auxiliary projections, is shown in Fig. 204. The first projection is made from the left-hand elevation at right angles to the centreline $C'D'$ of the branch pipe. The base line of the cone in the projection is drawn through the centrepoint E'' at right angles to the projection lines. The smaller end of the cone is drawn through the centrepoint F'' at right angles to the projection lines, and at a perpendicular distance from the

Auxiliary and Double Projections

base line equal to the distance *AB* taken from the right-hand elevation between the two ends. The diameter at each end and the sides of the cone are then drawn in.

The next step is to locate the position of the branch pipe in the first projection. In the right-hand elevation, the point *D* on the centreline of the branch pipe lies also on the line at right angles to

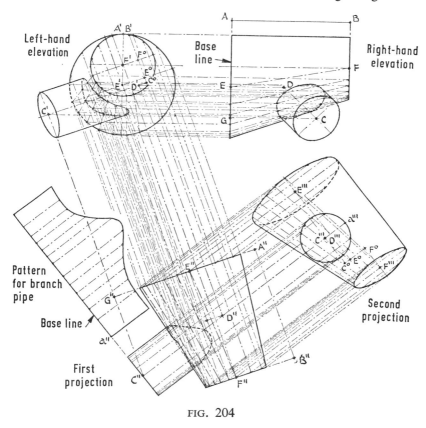

FIG. 204

the base of the cone from point *E*, the centre of the base. In the left-hand elevation, point *D'* lies to the right of point *E'* on the horizontal centreline of the base circle. Then in the first projection, the point *D"*, projected from *D'*, lies on a line perpendicular to the base line through *E"*, at a distance from *F"* equal to the length *ED* in the right-hand elevation. The point *C'* in the left-hand elevation is now projected into the first projection. Then from the right-hand elevation the perpendicular distance from the base is taken from point G to point C and marked off in the first projection from the extended base line at point G'' to point C''. A line drawn from C'' to D'' represents the true centreline of the branch pipe, and is lying flat in the plane of the paper. The end of the pipe now presents a straight line through C'' at right angles to $C''D''$.

Before the joint line can be determined a second projection is made in the direction of the centreline $C''D''$, or in other words the second projection is a view looking along the axis of the branch pipe. In that view the branch pipe presents a circle, and the two ends of the cone present ellipses.

It is assumed that the reader, at this stage, would have no difficulty in projecting the ellipses from the first projection. The most important point to observe in this part of the solution is the relative positions of the centres of the ellipses and the circle in the second projection.

Return now to the left-hand elevation and observe that the perpendicular distances of the two centrepoints E' and F' from the centreline of the branch pipe are equal to C^0E^0 and C^0F^0 respectively. In the second projection the centrelines through points C''', E''' and F''' are drawn at right angles to the projection lines, and the distances between them are made equal to C^0E^0 and E^0F^0 taken from the left-hand elevation. This is shown by the similar symbols C^0, E^0 and F^0 used in the second projection.

In the left-hand elevation, the bottom left-hand quadrant on the base circle and the corresponding quadrant on the top circle are now each divided into six equal parts, and lines drawn between the corresponding points on the top and bottom. These six points on the quadrants are next projected into the first projection to obtain the corresponding points on the base line through E'' and also on the top edge through F''. Lines are again drawn between the corresponding points on the two edges.

Next, the six points on both edges are projected into the second projection as far as the corresponding quarter ellipses, and the points joined as in the previous views.

Note. It is felt that detailed instructions for these operations are not necessary for readers at this stage of progress, and that general directions towards the solution will be readily followed and understood.

The next series of operations occur in the reverse order from the second to the first projection, then to the left-hand and right-hand elevations. In the second projection, it will be observed that the lines joining the two quarter ellipses cross the circle at various points. Those points on the circle are now projected back to the

Auxiliary and Double Projections

first projection to meet the corresponding lines on the cone. Points are thereby afforded through which the shape of the joint line is drawn in that view. Also, the lines just drawn into the first projection are continued onward to the end of the pipe at C''.

Next, the points on the joint line in the first projection are projected back to the left-hand elevation to meet the corresponding lines on the cone. Again, points are thereby afforded through which the shape of the joint line is drawn in that view. In this case careful drawing is needed in order to ensure that the joint line is plotted satisfactorily, as the slightest deviation would lead to considerable inaccuracy in the shape of the joint line.

Finally, the points on the joint line in the left-hand elevation are projected into the right-hand elevation to meet the corresponding lines on the cone, whereby points are again afforded through which the shape of the joint line is drawn in that view.

It will no doubt be appreciated that as the pattern for the branch pipe is unrolled from the first projection, the plotting of the joint lines in the left- and right-hand elevations constitutes additional exercises in projection.

The pattern for the branch pipe is unrolled at right angles to the centreline $C''D''$ in the first projection. The spacings along the base line are taken from the circle in the second projection, beginning at point a''' and marked off from point a'' in the pattern. From these spacings parallel lines are drawn at right angles to the base line. Then the distances from the end of the pipe at C'' to the points on the joint line are taken and marked up in their respective order from the base line in the pattern, thus affording points through which the joint curve is drawn.

9 Development of Complex Patterns and Spiral Chutes

The problems dealt with in this chapter do not necessarily require auxiliary projections for their solution, but most of them are somewhat out of the ordinary run of examples and need individual methods of approach to development. Some contain twisted or apparently twisted surfaces though the patterns in fact require only straightforward rolling to produce the desired shape of the surface. The spiral chutes belong to this class: when the segments are fitted together, the chute spirals round a centre column, either real or imaginary, but each segment is formed by simple rolling.

TRANSITION PIECE FROM SQUARE TO RECTANGLE

The transition piece shown in Fig. 205 connects an 18 in. square in a vertical plane, to a rectangle 20 in. long by 6 in. wide which inclines at an angle of 30° to a plane at right angles to that of the square. The four plates or patterns which compose the transition piece form rolled surfaces when shaped correctly to fit together.

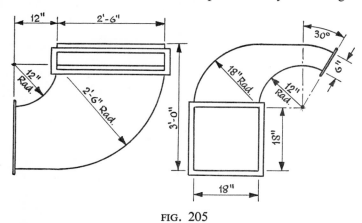

FIG. 205

Development of Complex Patterns and Spiral Chutes 255

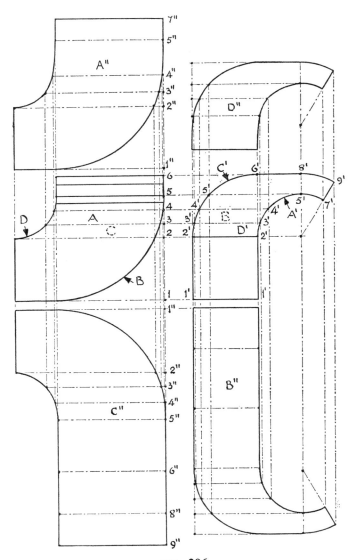

FIG. 206

There is no real twist in the sides of the transition piece, and the patterns may be shaped by ordinary rolling.

The development of the patterns is shown in Fig. 206. The front and side elevations are set out to the particulars given in Fig. 205, and a series of horizontal lines are drawn through both elevations as at 1, 2, 3, 4, 5 and 6. The four surfaces of the transition piece are lettered A, B, C and D. The front surface in the left-hand elevation is marked A, and in clockwise rotation the right-hand surface becomes B, the back surface is marked with the dotted letter C, and the left-hand surface is marked D.

Now taking the front surface A, imagine it to be flattened out as seen above it at A''. The lengths of the horizontal lines across the pattern will be the same as seen in the elevation below, but the height of the pattern, from point $1''$ to $7''$, will be equal to the length of the curve from $1'$ to $7'$ at A' in the right-hand elevation. Therefore the distances from $1'$ to $7'$ are taken from the curve A' and marked off up the vertical line from the edge of the front side A in the left-hand elevation, as at $1''-2''-3''-4''-5''-7''$. From these points parallel lines are drawn across horizontally to meet vertical lines drawn from the points on the curves in the elevation below. Thus points are obtained through which is drawn the contour of the pattern as shown at A''.

The pattern for C'' below the left-hand elevation is for the back surface C. The distances spaced off down the vertical line from the edge of the elevation, as from $1''$ to $9''$, are taken from the contour marked C' in the right-hand elevation. From these points parallel lines are drawn across horizontally to meet the lines drawn vertically downwards from the points on the curves in the elevation above. Points are thereby afforded through which is drawn the contour of the pattern for the back surface C.

The patterns for the two sides D and B are obtained from the right hand elevation, and are shown above and below at D'' and B''. The vertical spacings are taken this time from the left-hand elevation, and are equal to the distances round the curves D and B respectively. The 30° curved end piece at $5'7'9'8'$ presents its true shape and size in the right-hand elevation, and this is reproduced in the patterns as shown in the Figure at D'' and B''.

ANOTHER ROLLED-SURFACE TRANSITION PIECE

The transition piece shown in Fig. 207 represents a further example of a transition piece with rolled surfaces, in which the patterns may be shaped by simple rolling.

In this type of transition piece there is an important condition

Development of Complex Patterns and Spiral Chutes   257

which should be observed in relation to the design. In the left-hand elevation, Fig. 207, the corners joining the square and rectangle are represented by straight lines. In the right-hand elevation the throat and back are represented by the two quadrants of 9-in. and 30-in. radius respectively. From these two views the curves in the plan below the left-hand elevation are derived from the straight lines above and the quadrants in the right-hand elevation. The method

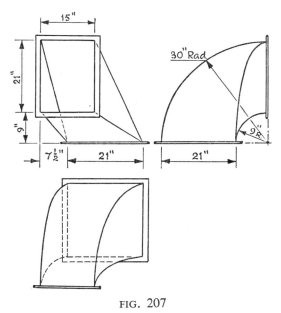

FIG. 207

of determining these curves, which represent the corners of the transition piece, is shown in Fig. 208, which at this stage should be self-explanatory.

The important condition referred to above relates to the view containing the straight lines, which represent the corners of the transition piece. In Fig. 207 the straight lines are contained in the left-hand elevation and the curves in the plan are obtained therefrom. In Fig. 210 it will be observed that the plan contains the straight lines and the left-hand elevation the curves. In this case the curves in the left-hand elevations are derived from the right-hand elevation and the plan. The method of determining these curves is shown in Fig. 211, and, as before, should be self-explanatory.

The difference in the treatment shown in these two examples is important as the choice applies to many types of transition piece of this kind. The basic principle underlying the setting out of the views is that one elevation, showing the throat and back, should conform to curves of definite radii, as in the present cases of 9-in. and 30-in. radii. Then either the plan or the other elevation may contain the straight lines representing the corners of the transition piece, but both of these views cannot contain the straight lines. To whichever view the straight lines are designated, the remaining view must contain the corresponding curves.

It is purely a matter of choice which of these methods be adopted, but the determination of the patterns must follow the method

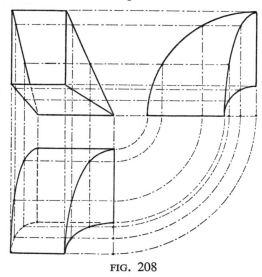

FIG. 208

decided upon. Although the square and the rectangle in both cases are the same size and in the same relative positions, the patterns differ according to the method of construction. This will be seen by an inspection of Figs. 209 and 212.

The patterns for the transition piece in Fig. 207 are shown developed in Fig. 209. The left- and right-hand elevations are first set out, and a vertical line drawn through the corner point 1′ in the left-hand elevation. The back quadrant in the right-hand elevation is next divided into three equal parts (or any other number according to convenience), and the points numbered 1, 2, 3 and 4. Horizontal lines are drawn through points 2 and 3 to cut both elevations. The throat quadrant is next divided into three equal parts and the points numbered 5, 6, 7 and 8. Horizontal lines are also drawn through points 6 and 7 to cut both elevations.

Development of Complex Patterns and Spiral Chutes

To develop the pattern for the back panel A, the spacings 1,2, 2,3 and 3,4 are taken from the back quadrant and marked off up the vertical line through point 1'. The points thus marked off are numbered 1", 2", 3" and 4", and horizontal lines are drawn from them to meet a series of vertical lines drawn through points 2', 3' and 4' in

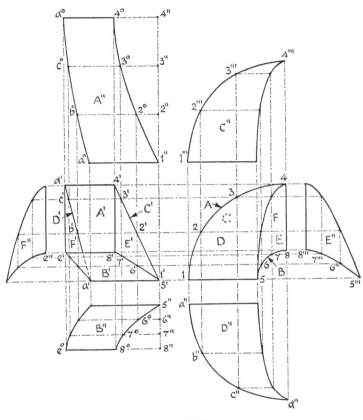

FIG. 209

the left-hand elevation. Points 2^0, 3^0 and 4^0 are thereby obtained on the curve of the pattern for the back panel. The corresponding curve on the opposite side of the panel is obtained by extending the horizontal lines through points 2^0, 3^0 and 4^0 to meet a series of vertical lines drawn from the opposite straight line in the left-hand elevation.

The pattern for the throat panel B is obtained in a similar manner by taking the spacings 5,6, 6,7 and 7,8 from the right-hand elevation, and marking them off down the vertical line through point $5'$ in the left-hand elevation. The points thus marked off are numbered $5''$, $6''$, $7''$ and $8''$, and from them are drawn horizontal lines to meet a series of vertical lines drawn through points $6'$, $7'$ and $8'$ in the left-hand elevation. Points 6^0, 7^0 and 8^0 are thereby obtained on the curve of the pattern for the throat panel. The corresponding curve

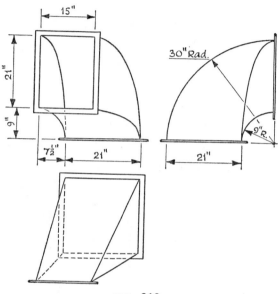

FIG. 210

on the opposite side of the panel is obtained by extending the horizontal lines through points 6^0, 7^0 and 8^0 to meet a series of vertical lines drawn from the opposite straight line in the left-hand elevation, to give points b^0, c^0 and d^0.

To develop the pattern for the cheek side C, the first step is to draw vertical lines through points 1, 2, 3 and 4, and also through points 5, 6, 7 and 8 in the right-hand elevation. A horizontal base line is then drawn through point $1''$ in any convenient position. It should be observed that the curve $1'''2'''3'''4'''$ in the cheek pattern must fit or match the curve $1''2^03^04^0$ in the pattern for the back panel, as these two edges are joined together in the course of fabrication. They must therefore be of the same length. To ensure this condition the spacings $1''2^0$, 2^03^0 and 3^04^0 are taken from the pattern for the

Development of Complex Patterns and Spiral Chutes 261

back panel and marked off across the vertical lines drawn from points 2, 3 and 4 in the right-hand elevation. Thus, in the cheek pattern C'', the distance $1'''2'''$ is equal to $1''2^0$, the distance $2'''3'''$ is equal to $2^0 3^0$ and $3'''4'''$ is equal to $3^0 4^0$. Now, from the two points $2'''$ and $3'''$, horizontal lines are drawn across to meet the vertical lines from points 6 and 7 in the right hand elevation. Points are thereby afforded through which the opposite curve is drawn in the pattern C''.

The pattern for the opposite cheek side D is obtained in a similar manner, and is shown below the right-hand elevation at D''. The vertical lines through the points 1, 2, 3 and 4, and also those through

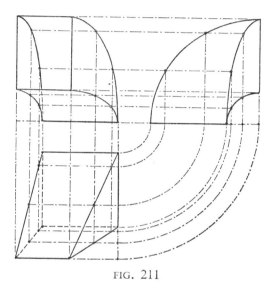

FIG. 211

points 5, 6, 7 and 8, are extended below the elevation, and a horizontal base line is drawn through point a'' in any convenient position. In this pattern the distances along the curve $a''b''c''d''$ must be equal to the distances along the curve $a^0 b^0 c^0 d^0$ in the pattern for the back panel. Therefore, the distances $a^0 b^0$, $b^0 c^0$ and $c^0 d^0$ are taken from the pattern for the back panel and marked off across the vertical lines drawn from points 2, 3 and 4 in the right-hand elevation, as shown in the Figure. Points are thereby afforded through which the curve $a''b''c''d''$ is drawn. Now, from the two points b'' and c'', horizontal lines are drawn across to meet the vertical lines from points 6 and 7 in the right-hand elevation. Thus points are afforded through which the opposite curve is drawn in pattern D''.

It remains now to develop the patterns for the two fillet pieces

E and F. These may best be seen in the left-hand elevation. Fillet E' lies between the points $4'$, $5'$ and $8'$. Fillet F' lies between d', a' and e'.

To develop the pattern for fillet E, the horizontal lines through points 1, 2, 3, 4 and 5, 6, 7, 8 are extended beyond the right-hand elevation, and a vertical base line drawn through point $8'''$ in any convenient position. In this pattern the distances along the curve $8'''7'''6'''5'''$ must be equal to the distances along the curve $8^06^05''$ in the pattern for the throat panel. Therefore the distances 8^07^0, 7^06^0 and $6^05''$ are taken from the throat panel and marked off across the horizontal lines drawn from points 7, 6 and 5 in the right-hand elevation as shown in the Figure. Points are thereby afforded through which the curve $8'''7'''6'''5'''$ is drawn. From the two points $7'''$ and $6'''$, vertical lines are now drawn to meet the horizontal lines from points 2 and 3 in the right-hand elevation. Thus points are afforded through which the opposite curve is drawn in pattern E''.

Finally, to develop the pattern for fillet F, the horizontal lines through points 1, 2, 3, 4 and 5, 6, 7, 8 are extended leftwards beyond the left-hand elevation, and a vertical base line drawn through point e'' in any convenient position. In this pattern the spacings along the inner curve from point e'' must be equal to the corresponding distances along the curve from point e^0 in the pattern for the throat panel. From the two middle points on the inner curve, vertical lines are drawn to meet the horizontal lines from points 2 and 3 in the right-hand elevation. Sufficient points are thereby afforded through which the two curves are drawn to complete pattern F''.

AN ALTERNATIVE TRANSITION PIECE

In Fig. 210, as mentioned above, it is the plan which contains the straight lines representing the corners of the transition piece. The method of determining the various curves is shown in Fig. 211, and as the procedure is similar to that of the previous example there should be little or no difficulty in following the construction from the illustration.

The patterns for the transition piece are shown developed in Fig. 212. When the three views are set out, the two quadrants in the right-hand elevation are each divided into three equal parts. Vertical and horizontal lines are then drawn through the points on the quadrants as in the previous example.

To develop the pattern for the back panel, the spacings 1–2–3–4 are taken from the right-hand elevation and marked off up the vertical line through $1'$ in the left-hand elevation, as shown at $1''$, $2''$, $3''$ and $4''$. Then horizontal lines are drawn from those points

Development of Complex Patterns and Spiral Chutes 263

to meet the corresponding vertical lines from points 2′, 3′, 4′ and b′, c′, d′ in the left-hand elevation. The two side curves are then drawn through the points thus obtained. It will be seen that these

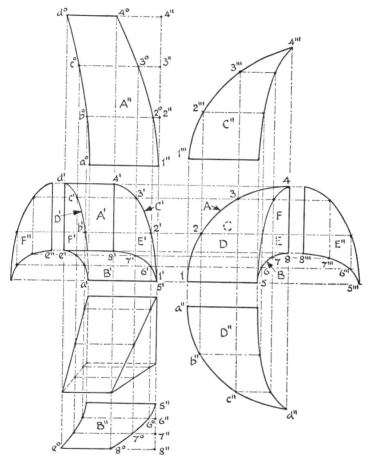

FIG. 212

two curves bend the opposite way to the corresponding curves in the previous example in Fig. 209.

The pattern for the throat panel is developed below the plan, and the method of procedure is basically the same as that for the throat panel in Fig. 209. The spacings 5–6–7–8 are taken from the smaller quadrant in the right-hand elevation and marked off down the vertical line through point 1' in the left-hand elevation as shown in the Figure. Horizontal lines are then drawn from these points to meet the corresponding vertical lines in points 6^0, 7^0 and 8^0, and also the lines from the opposite curve in the left-hand elevation. Points are thereby afforded through which the two curves are drawn to complete the pattern for the throat panel B''. Again it may be noted that the curves bend in the opposite way to the corresponding curves in the previous example.

The directions for developing the pattern for the side cheek C are precisely the same as for the corresponding pattern C in Fig. 209, and it may be noted that the shape of the pattern is similar but not exactly the same. The directions for developing the remaining patterns, D'', E'' and F'', are also the same as for the corresponding patterns in the previous example, and it may be noted that these, too, are similar in shape but are not exactly the same as those in Fig. 209. From this point, therefore, it is hoped that the reader will readily follow the solution of the remaining developments from the illustration.

A FLARED VENTILATOR HEAD

The example shown in Fig. 213 represents a ventilator head which fits on a square duct. From the square base the head flares to a

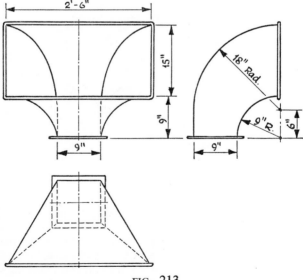

FIG. 213

Development of Complex Patterns and Spiral Chutes 265

larger rectangle in a plane at right angles to that of the square. Of the four sides which form the ventilator head, the back and throat panels are simple rolled surfaces, but the two side panels each contain a twist which will necessitate a certain amount of deformation

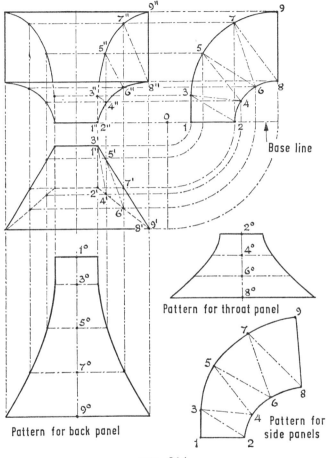

FIG. 214

in order to make them conform to the correct shape. In this example the side view contains the curves of definite radii and the plan view presents the straight lines representing the corners of the transition piece. The method of plotting the curves in the left-hand elevation and the development of the patterns are shown in Fig. 214.

In the first step, the back quadrant is divided into three equal parts and the points are numbered 1, 3, 5, 7 and 9 as shown in the Figure. Next, the throat quadrant is divided into three equal parts and the points numbered 2, 4, 6 and 8. Then all the points are dropped to the base line, and from point 0 on the base line the points are swung round through 90° to the vertical line through point 0. From there the lines are continued horizontally to cut the straight corner lines in the plan. From the corner lines the points are projected vertically upwards into the left-hand elevation to meet a series of horizontal lines drawn from the points on the quadrants in the right-hand elevation. Where the corresponding lines meet in the left-hand elevation, points are afforded through which the corner curves are drawn.

To develop the pattern for the back panel, the vertical centreline through the plan and the left-hand elevation is extended downwards, and the divisions 1–3–5–7–9 are taken from the back quadrant in the right-hand elevation and marked off down the centreline below the plan. The points are then numbered 1^0, 3^0, 5^0, 7^0 and 9^0. Horizontal lines are next drawn through the points thus marked off to meet a series of lines dropped vertically from the corner line $1'3'5'7'9'$, and also from the corresponding points on the opposite corner line in the plan. Points are thereby obtained through which the two curves are drawn in the pattern below the plan.

To develop the pattern for the throat panel, a vertical centreline is drawn in any convenient position, and the spacings 2–4–6–8 are taken from the throat quadrant in the right-hand elevation and marked off down the centreline. These points are numbered 2^0, 4^0, 6^0 and 8^0. Horizontal lines are next drawn through the points thus numbered. Then, from the plan, the widths are taken from the centreline to the points $2'$, $4'$, $6'$ and $8'$, and marked off on the corresponding lines on both sides of the centreline in the pattern. Thus, points are obtained through which the two curves are drawn.

The two side panels require somewhat different treatment. These cannot be unrolled as parallel line patterns. They need to be developed by the method of triangulation. In the right-hand elevation the surface is divided into triangles as seen by the numbering 1,2,3,4,5,6,7,8,9. This numbering is repeated on the corresponding points in the plan, and again in the left-hand elevation. The pattern is shown developed in the bottom right-hand corner in Fig. 214, and it is assumed that the reader will be sufficiently well versed in the method of triangulation to be able to follow the solution without difficulty. It should be observed that the patterns for both side panels will be exactly the same.

Development of Complex Patterns and Spiral Chutes

AN ALTERNATIVE FLARED VENTILATOR HEAD

The ventilator head illustrated in Fig. 215 is a further example of the alternative placing of the straight lines representing the corners of the transition piece. In Fig. 213 the straight lines were located in the plan. In Fig. 215 the straight lines are located in the left-hand elevation. It should also be noted that the square and rectangle are the same size and in the same relative positions as those illustrated in Fig. 213. The right-hand elevation is also the same as in the previous example. The corner curves in the plan are dependent on

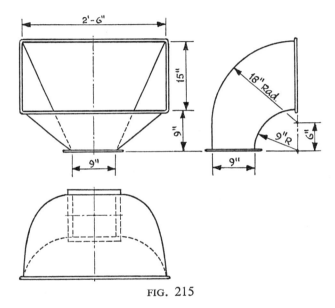

FIG. 215

the particulars given in the two elevations, and the method of determining the curves is similar to that applied in the previous example. But this difference of treatment in the design results in a marked difference in the shape of the patterns as will be seen by comparing those in Fig. 214 with those in Fig. 216.

The pattern for the back panel, Fig. 216, is obtained by extending the centreline from the left-hand elevation downwards through the plan, and marking off the spacings 1^0–3^0–5^0–7^0–9^0 taken from the back quadrant in the right-hand elevation. Horizontal lines drawn through those points to meet vertical lines dropped from points $1''$, $3''$, $5''$, $7''$ and $9''$, and similar points on the opposite side in the

268　*Sheet Metal Drawing and Pattern Development*

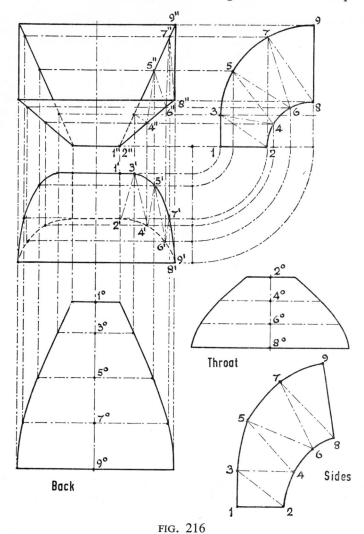

FIG. 216

left-hand elevation, afford points through which the curves of the back pattern are drawn.

Similarly, the pattern for the throat panel is obtained by drawing a centreline in any convenient position and marking off the spacings 2^0–4^0–6^0–8^0 taken from the throat quadrant in the right-hand elevation. Horizontal lines are drawn through those points, and the

Development of Complex Patterns and Spiral Chutes 269

distances in the left-hand elevation from the centreline to points 2″, 4″, 6″ and 8″ are taken and marked off on both sides of the centreline in the pattern. Points are thereby afforded through which the curves of the throat pattern are drawn.

The patterns for the side panels, which are both exactly the same, are developed by triangulation. The right-hand, left-hand and plan views are divided into triangles as shown in the Figure. Then by the use of the plan and the left-hand elevation, the development by triangulation should present a fairly straightforward exercise.

PREDETERMINED CURVES

In the example shown in Fig. 216A, the corners of the flared ventilator head are shown curved in both the plan and elevation. The curves in these two views are predetermined, and in this case the

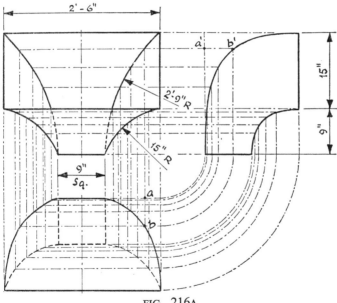

FIG. 216A

right-hand elevation does not, as in the two previous examples, contain curves of definite radii, but must conform to those set out in the plan and elevation.

In the side elevation the distances from the back of the head to the points on the curves must be the same as the corresponding distances in the plan from the back of the head to the points on the curves. For example, the distance *ab* in the plan corresponds to the distance *a′b′* in the side elevation. By following this rule, as shown by the construction in the Figure, the curves in the side elevation may be obtained, but do not conform to given radii.

From the series of examples given from Fig. 207 to Fig. 216A, the rule may be deduced that any two of the views may be drawn to a given specification, but the third view must conform to the conditions given in the other two. From this it may be seen that the best solution is to assign definite radii to the right-hand elevation, and straight line corners in either the plan or the front elevation. The method of developing the patterns for the ventilator head shown in Fig. 216A is precisely the same as that used in the two previous examples.

FLAT-SIDED CHEEKS

A further example is given in Fig. 216B, in which, instead of a throat radius of 9 in. in the side elevation, a straight line is drawn from *B* to *C*, and also from *A* to *C*. In the front elevation the line *AC* lies behind the line *BC*, thereby presenting the single line *A′C′–B′C′* which represents both.

By this construction each of the two side cheeks of the ventilator head may be made up of two flat pieces, *ACB* and *ACD*, with a bend along the line *AC*. The top curve in the side elevation from *A* to *D* presents a corresponding curve from *A′* to *D′* in the front elevation.

To determine the curve *A′D′* in the front elevation, the line *AC* in the side elevation is divided into a number of equal parts, in this case, six and lines are drawn vertically through them to meet the curve *AD* at the top and the base line *AE* at the bottom. From the points on *AC*, horizontal lines are drawn to cut the line *A′C′* in the front elevation, passing through to the corresponding line on the other side of the elevation. From the points obtained on *A′C′*, vertical lines are drawn upwards to meet horizontal lines drawn from the points on the curve *AD* in the side elevation. Points are thereby afforded through which the curves in the front elevation may be drawn.

A plan will be needed from which to develop the pattern for one of the side cheeks. To obtain the plan, the points on the line *A′C′* in the elevation are projected vertically, and a horizontal base line *MN* is drawn across them in any convenient position. Next, the six divisions are taken from the base line *AE* in the side elevation and marked down the centreline from *MN* in the plan. Horizontal lines are drawn parallel to *MN* through the points marked on the

Development of Complex Patterns and Spiral Chutes

centreline, to meet the vertical lines dropped from $A'C'$ in the elevation. A line drawn from A'' to C'' now represents the corner AC in the plan, and a line drawn from B'' to C'' represents the other corner BC. It will be noted that both of these are straight lines in the plan.

To develop the pattern for the cheek $ABCD$, lines are projected from the points on $A''C''$ in the plan at right angles to that line, and

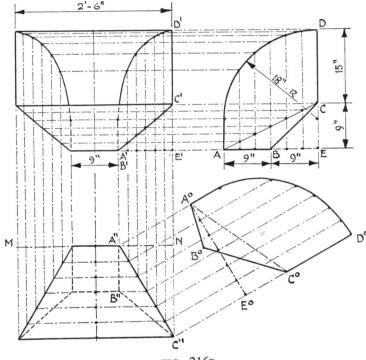

FIG. 216B

a datum line, A^0E^0 is drawn across them parallel to $A''C''$ in any convenient position. Next, the heights $E'-C'-D'$ are taken from the elevation and marked up the line in the pattern from E^0 to C^0 to D^0. Similarly, from the base line $A'E'$ in the elevation, the various heights are taken from the points thereon to the curve $A'D'$, and marked off from the datum line A^0E^0 on the corresponding lines in the pattern. Points are thereby afforded through which to draw the curve A^0D^0. A straight line drawn from A^0 to C^0 completes the portion $A^0C^0D^0$ which is a flat piece without a twist.

It remains now to add the bottom triangle $A^0B^0C^0$. The distance AB from the side elevation, or the distance $A''B''$ from the plan, both of which are true distances, may be taken and from point A^0 in the pattern an arc is drawn through point B^0. The next distance B^0C^0 is obtained by taking the plan distance $B''C''$ and triangulating it against the vertical height $E'C'$ in the elevation. The true diagonal is then taken and from C^0 in the pattern an arc is drawn to cut the previous arc in point B^0. This completes the triangle $A^0B^0C^0$ and the pattern for the cheek $A^0B^0C^0D^0$.

The patterns for the back and throat panels, it is assumed, should present no difficulty to the reader at this stage.

It may be well to note that, since the rectangular top is of the same size and is in the same relative position to the square bottom in each case, the examples given from Fig. 213 to Fig. 216B represent four different patterns for the same ventilator head.

A SPIRAL FINIAL

The finial shown in Fig. 217 is composed of twelve segments, which when rolled to shape and fitted together form the spiral ornamental finial as illustrated in the Figure.

In the plan the edges of the segments form a series of ovals on centrelines radiating at 30° from the centre of the circle which forms the outside circumference of the finial. The major axis of each oval is equal to the radius of the circle, and the minor axis is equal to the distance across its centre between the two adjacent 30° centrelines. This will best be seen by an inspection of the bottom oval in the figure. Since each segment finishes on the base of the finial, each of its edges in the plan represents only three-quarters of an oval. Thus, each oval begins at the centre of the circle and finishes on the dotted circle which represents the base of the finial in the plan.

One segment of the finial lies between two ovals in the plan. The segment selected for development in this case lies between the two ovals numbered 1 to 17 and 2 to 18. To plot the corresponding curves in the elevation, the vertical centreline of the finial is divided into a number of equal parts, in this case nine, and horizontal lines are drawn through them to cut the outside form of the finial. These lines may be regarded as horizontal cutting planes. To locate the cutting planes in the plan, the points where the horizontal lines cut the outside of the finial are dropped from the elevation to the horizontal centreline in the plan. Then, with centre the centre of the circle, and radii taken to the points obtained on the horizontal centreline, a series of inner circles are drawn. These circles represent the cutting planes in the plan.

Development of Complex Patterns and Spiral Chutes

Now, from the points where the inner circles cut the two selected ovals, lines are drawn into the elevation to meet the corresponding cutting planes. Points are thereby afforded through which the spiral curves are drawn. Fig. 217, only one set of lines, those from the

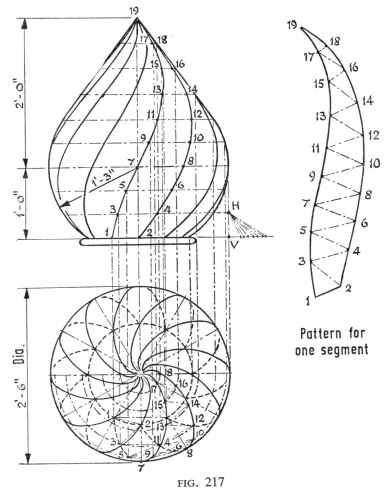

FIG. 217

bottom oval, are shown in order to avoid congestion of lines. It will of course be appreciated that the necessary points in the elevation can be plotted without actually drawing in the lines. By following the same method, all the remaining spiral curves may be plotted from the corresponding ovals in the plan.

To develop the pattern for the selected segment, the corresponding points on the two ovals in the plan are joined across from one to the other, as from 1 to 2, 3 to 4, 5 to 6, and so on. Then diagonals are drawn as from 2 to 3, 4 to 5, 6 to 7, and so on. Thus the segment is divided into triangles in the plan.

In triangulating these plan lengths, it should be noted that in the elevation the lines between points on the same cutting plane, as from 1 to 2, 3 to 4, 5 to 6, and so on, have no vertical heights. Therefore the true distances between these points for the pattern will be taken direct from the plan. All the diagonals between the the cutting planes, on the other hand, need triangulating against the vertical height. However, since the cutting planes in the elevation are all equally spaced, all these vertical heights are the same. Therefore only one vertical height need be used, as at *VH*.

From these observations the development of the pattern should be a matter of straightforward triangulation, and the specific details of procedure are not considered necessary for the reader at this stage.

A SPIRAL CHUTE

The spiral chute illustrated in Fig. 218 is similar to a spiral lobster-back pipe, the only difference being that the segments of which the chute is constructed are each only two-thirds of a complete cylindrical segment. This leaves one-third of the circumference of the segment open at the top. The illustration represents one revolution of the chute, but the chute may be extended to contain any number of revolutions if required.

Chutes of this kind are sometimes constructed to convey material vertically downwards at the angle of the spiral from one floor to another through a height of many feet or several revolutions. Although the top of the chute is open, the material does not overflow. Even in the case of a choke at the bottom, the angle of repose of the material is usually such that the material comes to rest in the chute without overflowing.

In developing the pattern for one segment—all the segments are the same—it is a good plan to develop the full lobster-back segment, and then use the portion required for the chute. For example, the full pattern for this chute is shown developed in Fig. 220, but only the portion between points 1 and 18 is actually required.

The setting out of this problem for development is shown in Fig. 219. However, before dealing with the actual development of the pattern, there are one or two important points to consider in the preliminary setting out.

Development of Complex Patterns and Spiral Chutes

In this example the chute contains twenty-four segments per revolution. In the plan, Fig. 219, the quadrant AB represents one quarter revolution of the centreline. This is divided into six sections, each of 15°, which represent the plan centrelines of the segments.

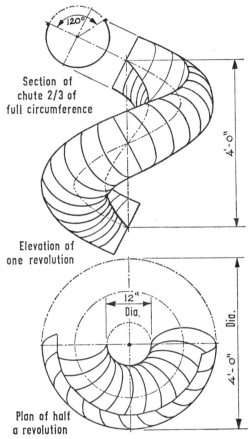

FIG. 218. *Spiral chute*

Next, the angle of inclination of the chute is set out in the elevation. This is determined by taking the length of the quadrant AB, or the six spacings along the curve between the 15-degree sections, and marking them off along the horizontal line through point A' to B'. Thus, $A'B'$ represents the length of one-quarter of a revolution in the plan. Now, from point B' in the elevation, the distance $B'C'$ is marked off at right angles to $A'B'$ and made equal to one-quarter

276 Sheet Metal Drawing and Pattern Development

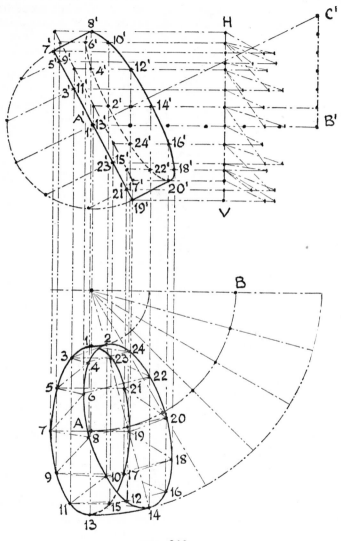

FIG. 219

Development of Complex Patterns and Spiral Chutes

of the pitch of the spiral. The pitch of the spiral is the distance between the points of one complete revolution, which in this case is given as 4 ft. Therefore $B'C'$ is made equal to 1 ft, and as there are six segments per quarter revolution, $B'C'$ is divided into six equal parts. Thus one of these parts represents the pitch of one segment. The line now drawn from A' to C' represents the slope or angle of inclination of the chute.

The joint line between the two segments adjacent to point A' will present a straight line through A' at right angles to $A'C'$. The length of that line is then made equal to the diameter of the chute, with A' as centre, as from 7' to 19'. Since the joint line is really a circle, it is next dropped into the plan as an ellipse in the usual manner by describing a semicircle on the diameter, dividing it into six equal parts, projecting the points back to the diameter, and dropping them into the plan to obtain the ellipse as shown on the axis 1 to 13 in the Figure.

It is now important to observe that each joint line between the segments presents a similar ellipse in the plan with its major axis along a 15-degree centreline. Therefore to set out one full segment in the plan, the ellipse obtained on the vertical centreline through point A is now repeated or swung round to the next 15-degree centreline, as shown in the Figure. Thus the two ellipses represent the joint lines or the edges on each side of one segment.

The next step is to determine the second ellipse in the elevation. This needs a little careful attention. The first joint line through the centrepoint A' is represented by a straight line at right angles to $A'C'$. The next joint line in the elevation, which corresponds to the second ellipse in the plan, will lie above the first joint line at a height equal to the pitch of one segment. Therefore, from the first point above B' a horizontal line is drawn back to the elevation. The centre of the second joint line, which in this case presents an ellipse, will lie on that horizontal line.

To plot the ellipse in the elevation, all the points on the second ellipse in the plan are projected vertically upwards into the elevation. The next manœuvre is to move all the points on the straight joint line at A' vertically upwards to the height of the pitch of one segment, and then horizontally to meet the corresponding lines from the second ellipse in the plan. This move is perhaps best seen by repeating the diameter at A', with all the points on it, at a height of one segmental pitch above A', as can be seen in Fig. 219. The points on the raised diameter are then projected horizontally to meet the corresponding vertical lines from the second ellipse in the plan. Points are thereby obtained through which the ellipse is drawn in the elevation.

The setting out of the plan and elevation is now sufficiently complete for the development of the pattern. The first important step is the numbering of the points in the plan and elevation. In the plan the numbering begins at the inside point on the first ellipse, numbered 1, and then to the corresponding inside point on the second

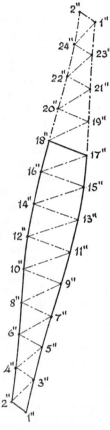

FIG. 220

ellipse, numbered 2. Following the corresponding points around the two ellipses, the numbers proceed in consecutive order to 23 and 24 and then back to 1 and 2. In the elevation the same points are numbered from 1' to 24' and back to 1'.

Having carefully numbered the points, the development of the pattern by triangulation is a straightforward process which should offer little difficulty if carefully followed. All the diagonals between the two ellipses in the plan must be triangulated against their appropriate vertical heights in order to obtain their true lengths, while all the spacings along both sides of the pattern must be one-twelfth of the circumference of the chute, or alternatively, taken from the spacings around the semicircle in the elevation. The pattern is shown developed in Fig. 220.

A SPIRAL CHUTE WITHOUT CENTRE COLUMN

The illustration in Fig. 221 represents a spiral chute similar to the previous example, but with the difference that whereas it is usual for a spiral chute to circulate round a central column, either real or imaginary, the present example has no such central column, and circulates around what might be regarded as a single vertical line. Also, the section of the chute is semicircular as shown at the top of the illustration. This example, like the previous one, represents one complete revolution, as from A to B in the elevation. In the plan the segments form half ellipses, beginning at the top section at A, as shown in the plan by the somewhat heavier line. One revolution of the chute is composed of 24 segments.

The method of setting out the problem for development is similar to that of the previous example. The quadrant AB in the plan represents one quarter revolution of the centreline, and is divided into six sections of 15 degrees each.

The angle of inclination of the chute is obtained by taking the six spacings from the quadrant and marking them off along the horizontal line through point A' to B' in the elevation. Then along the vertical line from B', the distance $B'C'$ is made equal to one-quarter of the pitch, which is the distance between A and B in Fig. 221. The height $B'C'$ is then divided into six equal parts, each of which represents the pitch of one segment. The line now drawn from A' to C' represents the slope or angle of inclination of the chute.

The straight joint line through point A' is now drawn at right angles to $A'C'$ and made equal to the diameter of the chute, as from point 7' to 19'. This joint line is then dropped into the plan to obtain the first ellipse on the vertical centreline through point A. The second ellipse is next obtained by repeating the first ellipse on the adjacent 15-degree centreline.

As in the previous example, the ellipse in the elevation is determined by projecting the points on the second ellipse in the plan vertically upwards into the elevation, raising the straight joint line 7'19' vertically upwards through a height equal to the pitch of one segment, and then projecting the points on the raised line horizontally to meet the corresponding vertical lines from the second ellipse in the plan.

The next important step is the numbering of the points in the plan and elevation prior to the development of the pattern. The

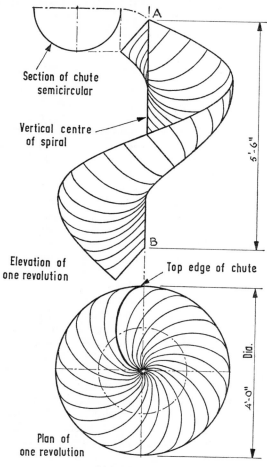

FIG. 221. *Spiral chute*

method of numbering is the same in this case as in the previous example shown in Fig. 219. It should be noted, however, that as there is no radius in the plan to represent a central column, the point 2 lies vertically above point 1, so that points 1 and 2 are represented by a single point in the plan.

Development of Complex Patterns and Spiral Chutes

In this example, although the numbering is carried all round the segment, as from 1 to 24, only the bottom half is shown developed (Fig. 222). The procedure of development of the pattern is now left to the reader, as the method is straightforward triangulation.

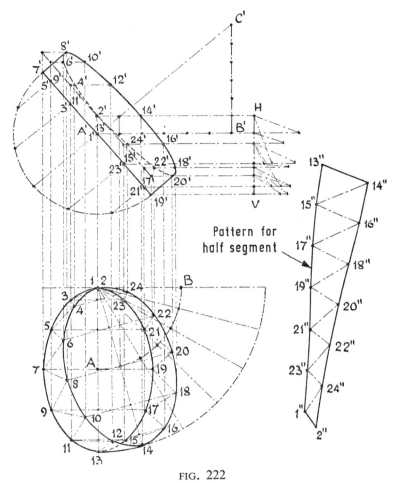

FIG. 222

A DIVIDED FEED HOPPER

The hopper shown in Fig. 223 was originally designed to fit under the base of a grading machine by which material was divided to feed two rotary grinders. The vertical distance allowable between the square at the top and the two circles at the bottom was somewhat limited. The front of the hopper was required to be open for accessibility by the attendant for inspection and maintaining a satisfactory flow of material.

This example is presented as a further application of the square-to-circle transforming piece. One half of the hopper is shown developed in Fig. 224. While the design of the hopper is perhaps

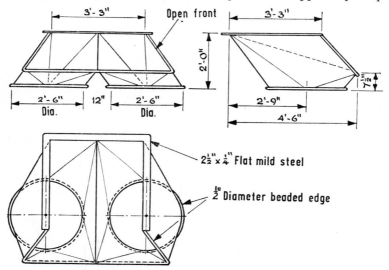

FIG. 223. *Divided feed hopper from grading machine to two rotary grinders*

a little unusual, the development of the pattern does not contain anything new in the matter of procedure, and is given as an additional exercise in development by triangulation.

When the two parts of the hopper are assembled, the two edges along the side numbered 3 to 7 come together to form a sloping ridge between the two halves. The three-sided flat mild steel frame at the top affords a means of fixing the hopper to the grading machine, and the beaded edges at the bottom, while not actually points of fixture, form fairly close contacts with the tops of the grinders.

GUSSET PIECE FOR OFF-CENTRE UNEQUAL-DIAMETER PIPES

The gusset piece between the branch and the main pipe shown in Fig. 225 is presented as a variation from the more usual example of

Development of Complex Patterns and Spiral Chutes

a gusset between pipes of equal diameters. The development of this gusset piece is shown in Fig. 226.

The left-hand elevation with the branch at 55° is first set out, and the semicircle drawn on the end of the branch pipe. The end of

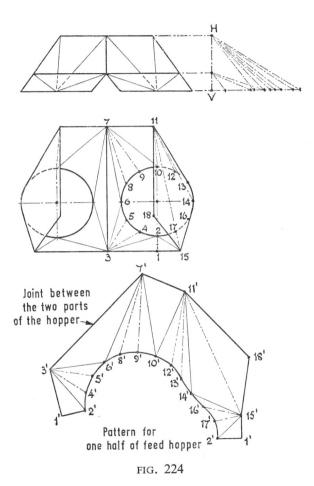

FIG. 224

the branch pipe is then projected into the right-hand elevation to obtain the ellipse as shown in that view. The points on the ellipse are then projected upwards to intersect the circle which represents the cross-section of the larger pipe. The right-hand side of the branch pipe is drawn tangentially to the right-hand side of the circle, as shown. From the points on the circle where the branch pipe intersects, lines are drawn back to the left-hand elevation to meet corresponding lines drawn from the end of the branch pipe. Points are thereby obtained through which is drawn the shape of the joint line between the two pipes.

Next, the cross-section of the gusset piece is projected as shown in the Figure, and the semicircular bottom is divided into the usual six equal parts and the points projected back to the elevation to cross the corresponding lines on the branch pipe. The joint line

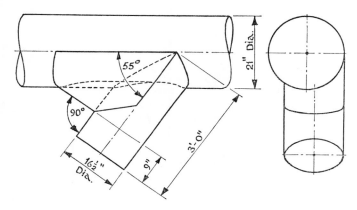

FIG. 225. *Gusset piece for off-centre unequal-diameter pipes*

between the bottom of the gusset piece and the branch pipe is thereby obtained as the straight line through the points of intersection. The curved end at the top of the gusset piece is obtained by extending the horizontal lines from the circle in the right-hand elevation to meet the corresponding lines projected from the cross-section of the gusset piece. The remainder of the gusset piece in the left-hand elevation is completed by drawing the two triangles *ABC* at the front and *DEC* at the back.

The pattern for the gusset piece is "unrolled" in the usual way, and is shown projected above the elevation in the Figure. It is assumed that the method of procedure will be clear from the illustration and will present no difficulty to the reader as it is based on simple parallel line development.

Development of Complex Patterns and Spiral Chutes 285

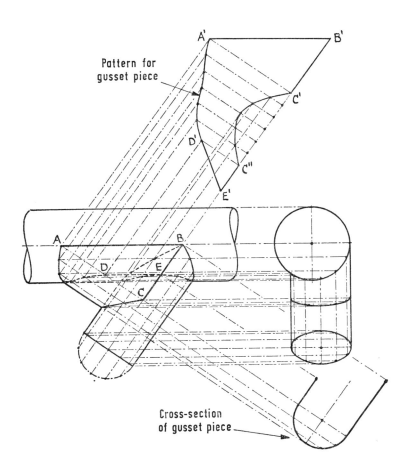

FIG. 226

10 Duct Layout with Developments

Since many of the intricate problems of pattern development are associated with duct or pipework, this chapter is devoted to a number of examples which illustrate the setting out of sections of ductwork and the development of the patterns for some of the units included in the sections. In general, the sections dealt with are simple arrangements of pipework, with branches, hoods and hoppers, and as these units often contain some special point of geometrical or constructional interest, the pattern developments are described in detail. Sometimes several alternative methods of construction may be envisaged to produce the same unit, as will be seen by reference to the branch piece shown in Fig. 230 (p. 292). These involve differences in geometrical conception which in turn must influence the methods of development of the patterns.

Fume Hoods and Extraction Ducts

The fume hoods and extraction ducts shown in Fig. 227 represent a typical section of ductwork comprising, apart from the lobster-back bends, a branch piece and a hood which require pattern development.

THE HOOD

Sometimes difficulties arising in the construction of an article may be due to the elusive character of pattern drafting. The hood illustrated in Fig. 227 is a typical example of draughtsmanship, sound in geometrical conception, yet two different patterns may be developed to make the hood exactly as shown. The important condition is, however, that the hood must be worked up in accordance with the method of development, otherwise the correct form cannot be arrived at.

Duct Layout with Developments

The point of difficulty occurs on the sides or cheeks of the hood, as at *MNOP* in the left-hand hood in the side elevation. It will be seen that there are four straight lines, *MN*, *NO*, *OP* and *PM*, enclosing the quadrilateral *MNOP*. The point is that the metal inside *MNOP* is not flat or in one plane and indeed the hood cannot be made with this portion flat, though from the drawing it might

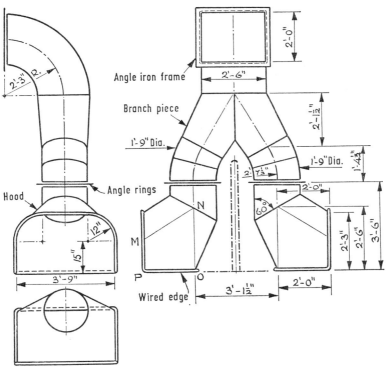

FIG. 227. *Fume hoods and extraction ducts*

appear that it could be. There must be a bend or kink across one of its diagonals as shown at *A* or *B* in Fig. 228. A bend across either diagonal will enable the metal to assume the correct form as shown in Fig. 227, but with the difference that a bend across the longer diagonal at *A* will be inwards, or "knuckle in," while a bend across the shorter diagonal at *B* will be outwards, or "knuckle out."

The patterns developed to these two alternatives will be appreciably different, though they may both be worked up to meet the conditions shown in the drawing. The two patterns are shown in Fig. 229, and it will be observed that the angle 3″7″8″ in the Figure at (*a*) is not a right angle. It is, in fact, about 96°, whereas in the pattern at (*b*) the corresponding angle 3″7″8″ is 90°. Also, the straight cross-distance between points 3″ and 8″ is greater in (*a*) than in (*b*). Moreover, the difference in the angle 3″7″8″ also affects to some extent the shape of the curve between points 8″ and 13″.

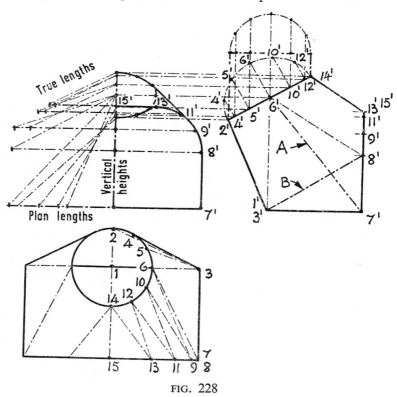

FIG. 228

These differences in the two patterns emphasize the need for care in the process of dividing up the surface for development by triangulation, and the importance of subsequently shaping the article in accordance with the method of surface division.

In dividing up the surface of this hood for triangulation, beginning at the centre of the back and numbering one half only, the first portion from 1 to 6 is straightforward tallboy triangulation. Then comes the quadrilateral 6,8,7,3. Perhaps the better choice of diagonal

Duct Layout with Developments

would be from 3 to 8, since this, as the bend is "knuckle outwards," would give the hood a little greater capacity than the alternative diagonal. In this construction the triangle 3,7,8 falls in a vertical plane. The next section, from point 8 to point 14, is simple triangulation between the two curved edges comprising points 8,9,11,

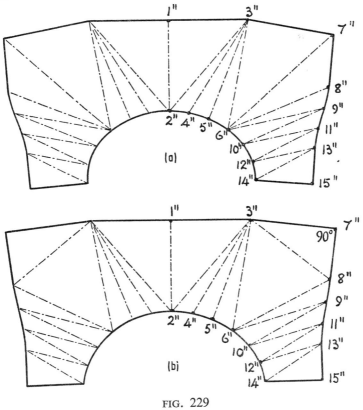

FIG. 229

13 on the front of the hood and points 6,10,12,14 on the joint edge at the top. Finally, the triangle 13,14,15 forms a flat piece on the front of the hood.

Prior to developing the pattern, a vertical height line is first erected and all the points in the elevation projected to it. In Fig. 228 only one half of the front elevation is shown, and the centreline of this view is used as the vertical height line. Then, since the true shape of the joint line between points 2 and 14 is elliptical, half of the true ellipse is set out as shown in the side elevation on the axis

2′14′. The spacings around this elliptical curve then give the true distances between the points 2′, 4′, 5′, 6′, 10′, 12′ and 14′, which will be required for the pattern.

To develop the pattern, as illustrated in Fig. 229(b) the true distance 1′ to 2′ is taken direct from the side elevation and marked off as from 1″ to 2″ in the pattern in any convenient position. Next, the true distance 1,3 is taken direct from the plan, and from point 1″ in the pattern an arc is drawn through point 3″. Now the plan length 2,3 is taken and marked off along the base line at right angles to the vertical height. The true length diagonal is then taken up to the point level with 2′, and from point 1″ in the pattern an arc is described to cut the previous arc in point 3″. This completes the first triangle.

For the second triangle, the plan length 3,4 is taken and marked off along the base line at right angles to the vertical height. The true length diagonal is taken up to the point level with 4′, and from point 3″ in the pattern an arc is drawn through point 4″. Next, the true distance is taken corresponding to 2′4′ round the ellipse in the side elevation, and from point 2″ in the pattern an arc is described cutting the previous arc in point 4″.

For the third triangle, the plan length 3,5 is taken and marked off at right angles to the vertical height. The true length diagonal is then taken up to the point level with 5′, and from point 3″ in the pattern an arc is drawn through point 5″. The true distance corresponding to 4′5′ round the curve of the ellipse in the side elevation is now taken, and from point 4″ in the pattern an arc is described cutting the previous arc in point 5″.

For the next triangle, the plan length 3,6 is taken and marked off at right angles to the vertical height. The true length diagonal is then taken up to the point level with 6′, and from point 3″ in the pattern an arc is drawn through point 6″. Now, the true distance is taken corresponding to 5′6′ round the curve of the ellipse in the side elevation, and from point 5″ in the pattern an arc is drawn cutting the previous arc in point 6″.

The next triangle, 3″6″8″, is obtained by taking the plan length 6,8 and marking this distance off at right angles to the vertical height, which is not taken from the bottom base line, but from a new base line level with point 8′. Next, the true length diagonal is taken up to the point level with 6′, and from point 6″ in the pattern an arc is drawn through point 8″. It should be observed that in the elevation the distance between 3′ and 8′ represents its true length. Therefore the distance 3′8′ is taken direct from the elevation, and from point 3″ in the pattern an arc is described cutting the previous arc in point 8″.

Duct Layout with Developments

Since the triangle 3′7′8′ is in a vertical plane, the view in the side elevation presents its true size. Therefore this triangle is added to the pattern by taking the true distance 3′7′ direct from the side elevation, and from point 3″ in the pattern an arc is drawn through point 7″. Then the true distance 8′7′ is taken direct from the elevation, and from point 8″ in the pattern an arc is described cutting the previous arc in point 7″.

For the next triangle, the plan length 6,9 is taken and marked off at right angles to the vertical height, but again it will be noted that its base line must be level with point 9′. Then the true length diagonal is taken up to the point level with 6′, and from point 6″ in the pattern an arc is drawn through point 9″. Now the true distance 8′9′ is taken direct from the front elevation, and from point 8″ in the pattern an arc is described cutting the previous arc in point 9″.

For the next triangle, the plan length 9,10 is taken and marked off at right angles to the vertical height along the base line level with point 9′. The true length diagonal is taken up to the point level with 10′, and from point 9″ in the pattern an arc is drawn through point 10″. The true spacing 6′10′ is now taken from the curve of the ellipse in the side elevation, and from point 6″ in the pattern an arc is described cutting the previous arc in point 10″.

From this point the remainder of the pattern should be readily followed from the directions given above. The chief points to observe are that in triangulating the zigzag diagonals 10,11, 11,12, 12,13, 13,14 and 14,15, these should be marked off at right angles to the vertical heights along the appropriate base lines. The true spacings for the top edge, 10′12′ and 12′14′, should be taken from the curve of the ellipse in the side elevation, and the true spacings for the bottom edge, 9′11′, 11′13′ and 13′15′, should be taken direct from the front elevation.

THE BRANCH PIECE

Referring now to the second problem, the branch piece shown in Fig. 227, this also may be developed in a number of different ways, resulting in a series of different patterns, all of which could be made to suit the requirements of the drawing. According to the details given, the branch piece is composed of two limbs which together transform from a rectangle at one end to two circles at the other. Various geometrical constructions could be applied to meet the conditions given, some of which are illustrated in Fig. 230. The differences lie on the inner part of the branch piece which connects the semicircle at the top to the joint line between the limbs. The form of the joint could be almost any reasonable shape which could be transformed or connected to the upper half of the circle at the top of the limb.

Perhaps the best construction is that shown at Fig. 230(d), which is selected for development in Fig. 231. The two limbs which form the branch piece are similar, so that the development for one will serve for both. The elevation and plan of the branch piece are shown in Fig. 231, where one limb is divided up for triangulation.

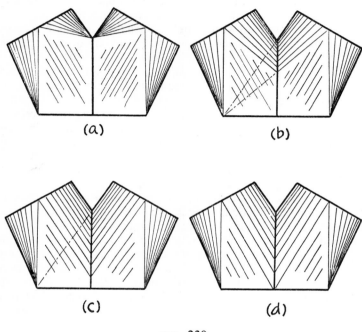

FIG. 230

It will be noted, perhaps more easily in the elevation, that the part numbered from 1′ to 7′ constitutes half of an ordinary tallboy transformer from rectangle to circle. The other part, between the two lines 6′7′ and 12′13′, transforms from a semi-ellipse at the joint line 7′12′ to a semicircle at the top, 6′13′. Half of the semi-ellipse at the joint line is shown turned back through 90° to present its true shape, as shown by the curve between points 7‴ and 12′.

Prior to developing the pattern a vertical height line VH is erected, and all the points in the elevation are projected horizontally to VH. The lines from 8′, 10′ and 12′ are extended beyond VH to serve as base lines in the process of triangulation.

Duct Layout with Developments

To develop the pattern, first the true distance 1′2′ is taken direct from the elevation and marked off in any convenient position, as at 1″2″, to begin the pattern. Next, the plan length 2,3 is taken and marked off along the bottom base line level with 3′. The true length

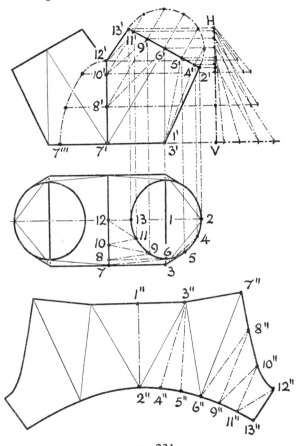

FIG. 231

diagonal is taken up to the point level with 2′, and from point 2″ in the pattern an arc is drawn through point 3″. The true distance 1,3 is now taken direct from the plan, and from point 1″ in the pattern an arc is described cutting the previous arc in point 3″. This completes the first triangle.

For the second triangle, the plan length 3,4 is taken and marked off along the bottom base line at right angles to the vertical height.

The true length diagonal is taken up to the point level with 4′, and from point 3″ in the pattern an arc is drawn through point 4″. Next, one of the equal spacings is taken from the circumference of the semicircle in the elevation, and from point 2″ in the pattern an arc is described cutting the previous arc in point 4″.

For the next triangle, the plan length 3,5 is taken and marked off along the bottom base line at right angles to the vertical height. The true length diagonal is taken up to the point level with 5′ and from point 3″ in the pattern an arc is drawn through point 5″. Then, another of the equal spacings is taken from the circumference of the semicircle in the elevation, and from point 4″ in the pattern an arc is described cutting the previous arc in point 5″.

For the next triangle, the plan length 3,6 is taken and marked off along the bottom base line at right angles to the vertical height. The true length diagonal is taken up to the point level with 6′ and from point 3″ in the pattern an arc is drawn through point 6″. Now, another of the equal spacings is taken from the circumference of the semicircle in the elevation, and from point 5″ in the pattern an arc is described cutting the previous arc in point 6″.

For the next triangle, the plan length 6,7 is taken and marked off along the bottom base line at right angles to the vertical height. The true length diagonal is taken up to the point level with 6′, and from point 6″ in the pattern an arc is drawn through point 7″. Now, the true distance 3,7 is taken direct from the plan, and from point 3″ in the pattern an arc is described cutting the previous arc in point 7″.

Next the plan length 6,8 is taken and marked off at right angles to the vertical height along the base line which, this time, is level with point 8′. The true length diagonal is taken up to the point level with 6′, and from point 6″ in the pattern an arc is drawn through point 8″. The first true spacing is now taken up the curve of the semi-ellipse in the elevation corresponding to 7′8′, i.e. the first spacing up the curve from point 7‴, and from point 7″ in the pattern an arc is described cutting the previous arc in point 8″.

Next the plan length 8,9 is taken and marked off along the base line level with point 8′. The true length diagonal is taken up to the point level with 9′ and from point 8″ in the pattern an arc is drawn through point 9″. Now, another of the equal spacings is taken from the semicircle in the elevation and from point 6″ in the pattern an arc is described cutting the previous arc in point 9″.

Next the plan length 9,10 is taken and marked off at right angles to the vertical height along the base line level with point 10′. The true length diagonal is taken up to the point level with 9′, and from point 9″ in the pattern an arc is drawn through point 10″. Then the

Duct Layout with Developments

second true spacing is taken up the curve of the semi-ellipse in the elevation, that corresponding to 8'10', and from point 8" in the pattern an arc is described cutting the previous arc in point 10".

The remaining three triangles needed to complete the pattern should readily be obtained by following the directions for the last three triangles, but applying them to the appropriate remaining points.

Dust and Fluff Extraction Ducts

One of the most important conditions on which the success of pattern development by triangulation depends is a correct interpretation of the geometrical form of the object to be developed. This particularly so when the object is complex in structure, that is, a combination of two or more geometrical types.

The arrangement drawing shown in Fig. 232 is a typical example of the draughtsmanship which finds its way to the sheet metal craftsman who has to make the hoppers and pipework. He also has to develop the patterns to produce the work to drawing, and sometimes extend on the work of the draughtsman in order to set out his requirements for the developments.

A COMPLEX HOPPER

The hopper shown at A, Fig. 232, is an example of complex geometrical construction which needs carefully analysing before applying the method of development. For instance, in the plan and elevation of this hopper shown in Fig. 233 it will be seen that the quarter circle in the plan between the points 13 and 19 forms the base of a right cylindrical portion standing vertically up to the oblique cut-off at the top between points 14 and 20. This cut-off is seen best in the elevation between points 14' and 20'. Since this portion of the hopper is cylindrical the parallel line method of development will be most appropriate.

The adjoining portion of the hopper between points 6 and 13 at the bottom, and between points 7 and 12 at the top, may at first sight appear to be a portion of an oblique cylinder leaning from the quarter circle 6 to 13 in the plan to the similar quarter circle 7 to 12 as seen also in the plan. But it will be observed that the edge between 7 and 12 also leans at the rather steep angle between 7' and 12' in the elevation. If this section of the hopper were a portion of an oblique cylinder, this edge would not present the quarter circle 7 to 12 as given in the plan, but would result in a curve as

296   Sheet Metal Drawing and Pattern Development

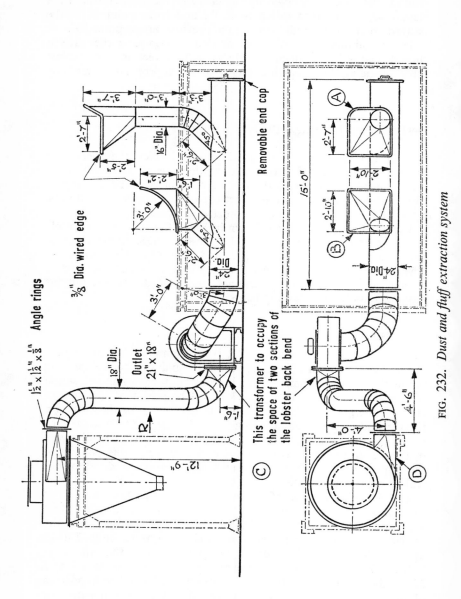

FIG. 232. *Dust and fluff extraction system*

Duct Layout with Developments

shown between points 7''' and 12''' in the section on the right-hand side of the plan in Fig. 233. Therefore, the portion of the hopper between 6 to 13 and 7 to 12 cannot be developed by the parallel line method, but must be dealt with by the method of triangulation.

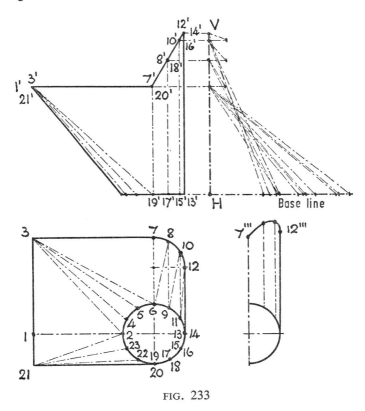

FIG. 233

The remaining part of the hopper between the semicircle 6,2,19 and the rectangle 7,3,21,20 may be developed by straightforward triangulation.

Prior to developing the pattern, divide the circle in the plan into twelve equal parts as from 2 to 23, and the quadrant 7 to 12 into three equal parts. In this development the joint is arranged to occur on the front flat side as from 1 to 2. Divide the surface of the hopper into triangles as shown in the plan in Fig. 233. Next erect a vertical height line *VH* in the elevation. Extend a base line from point H, and project the points on the sloping edge 7' to 12' horizontally to the vertical height line.

To develop the pattern as shown in Fig. 234, the plan length 1,2 is taken and marked off from point H along the base line at right angles to the vertical height. The true length diagonal is taken up to the point level with 1', and marked off in any convenient position, as from 1" to 2", to begin the pattern. Next, the plan length 2,3 is taken and marked off from point H along the base line. The true length diagonal is taken up to the point level with 3', and from point 2" in the pattern an arc is drawn through point 3". Now, the

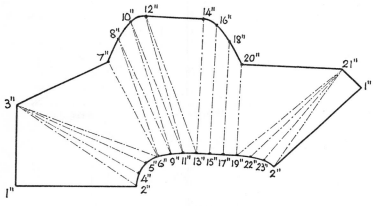

FIG. 234

true distance 1,3 is taken direct from the plan and from point 1" in the pattern an arc is described cutting the previous arc in point 3". This completes the first triangle.

For the second triangle, the plan length 3,4 is taken and marked off from point H along the base line. The true length diagonal is taken up to the point level with 3', and from point 3" in the pattern an arc is drawn through point 4". Next the true spacing 2,4 is taken direct from the plan, and from point 2" in the pattern an arc is described cutting the previous arc in point 4".

For the third triangle, the plan length 3,5 is taken and marked off along the base line. The true length diagonal is taken up to the point level with 3', and from point 3" in the pattern an arc is drawn through point 5". Next, the true spacing 4,5 is taken direct from the plan, and from point 4" in the pattern an arc is described cutting the previous arc in point 5".

Duct Layout with Developments

For the next triangle, the plan length 3,6 is taken and marked off along the base line. The true length diagonal is taken up to the point level with 3′, and from point 3″ in the pattern an arc is drawn through point 6″. Next the true spacing 5,6 is taken direct from the plan, and from point 5″ in the pattern an arc is described cutting the previous arc in point 6″.

For the next triangle, the plan length 6,7 is taken and marked off along the base line. The true length diagonal is taken up to the point level with 7′, and from point 6″ in the pattern an arc is drawn through point 7″. Next, the true distance 3,7 is taken direct from the plan, and from point 3″ in the pattern an arc is described cutting the previous arc in point 7″.

For the next triangle, the plan length 6,8 is taken and marked off along the base line. The true length diagonal is taken, this time up to the point level with 8′, the first point up the slope in the elevation. Then from point 6″ in the pattern an arc is drawn through point 8″. The next step is an important one. The true distance has to be obtained between points 7 and 8, and as the spacing in the plan is not a true distance and the distance between 7′ and 8′ in the elevation is not a true length, the plan distance must be triangulated against the vertical height in order to obtain the true diagonal. Therefore the plan length 7,8 is taken and marked off from the vertical height line along the base line projected from point 7′ in the elevation. The true length diagonal is then taken up to the point level with 8′, and from point 7″ in the pattern an arc is described cutting the previous arc in point 8″.

From this point the method of triangulation should readily be followed for the next six triangles. Care should be taken, however, to see that the true distances between points 8 and 10 and between points 10 and 12 are obtained by triangulating the plan lengths against the appropriate vertical heights in the elevation.

Assuming now that the pattern has reached the line 13″14″, the next portion will be that of the cylindrical section from 13,14 to 19,20. To obtain this section of the pattern, draw the line 13″19″ at right angles to 13″14″, and mark off the three spacings 13″15″, 15″17″ and 17″19″ equal to the corresponding spacings between the points 13, 15, 17 and 19 in the plan. Next draw the lines 15″16″, 17″18″ and 19″20″ all parallel to 13″14″. Now take the true heights 15′16′, 17′18′ and 19′20′ direct from the elevation and mark them off in the pattern from points 15″, 17″ and 19″. This completes the parallel line portion of the pattern.

The remainder of the pattern is straightforward triangulation which should be readily developed by following the trend of the directions given for the first part of the pattern.

A HOPPER WITH EXTENDED SIDES

Referring now to the arrangement drawing in Fig. 232, it will be seen that the second hopper at *B* is not so complex in structure as that at *A*, apart from the addition of the back and two sides at the top of the hopper. It may be of interest to develop the pattern in one piece as shown in Fig. 236.

The plan and elevation are shown in Fig. 235. Geometrically, the body of the hopper is a tallboy transformer, from a square at

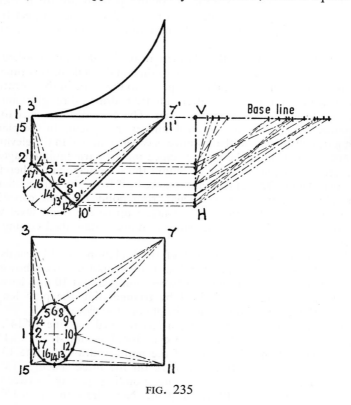

FIG. 235

the top to a circle at the bottom. The circle at the bottom is inclined at an angle of 45°, thus being seen as an ellipse in the plan. The joint is arranged vertically on the front of the hopper, as from 1 to 2. The usual method of dividing a tallboy transformer into triangles is adopted for development, and the points numbered accordingly as shown in the plan.

Duct Layout with Developments 301

Prior to developing the pattern, a vertical height line is erected as at *VH*. It will be seen that the vertical height line is inverted in Fig. 235, and that the base line is at the top instead of at the bottom. This, of course, makes no difference to the method of finding true length diagonals, which merely occur below instead of above the base line in this example.

To develop the pattern, the true vertical distance 1′2′ is taken direct from the elevation and marked off in any convenient position for the first line 1″2″ in the pattern. Next, the plan length 2,3 is taken, and marked off from *V* along the base line at right angles to *VH*. The true length diagonal is taken down to the point level with 2′, and from point 2″ in the pattern an arc is drawn through point 3″. The true distance 1,3 is now taken direct from the plan, and from point 1″ in the pattern an arc is described cutting the previous arc in point 3″. This completes the first triangle.

For the second triangle, the plan length 3,4 is taken and marked off from *V* along the base line. The true length diagonal is taken down to the point level with 4′, and from point 3″ in the pattern an arc is drawn through point 4″. Next, one of the equal spacings is taken from the projected semicircle on the line 2′10′ in the elevation, and from point 2″ in the pattern an arc is described cutting the previous arc in point 4″.

For the third triangle, the plan length 3,5 is taken and marked off along the base line. The true length diagonal is taken down to the point level with 5′, and from point 3″ in the pattern an arc is drawn through point 5″. Next, one of the equal spacings is taken from the semicircle in the elevation, and from point 4″ in the pattern an arc is described cutting the previous arc in point 5″.

For the next triangle, the plan length 3,6 is taken and marked off along the base line. The true length diagonal is taken down to the point level with 6′ and from point 3″ in the pattern an arc is drawn through point 6″. Then, one of the equal spacings is taken from the semicircle in the elevation, and from point 5″ in the pattern an arc is described cutting the previous arc in point 6″.

For the next triangle, the plan length 6,7 is taken and marked off along the base line. The true length diagonal is taken down to the point level with 6′, and from point 6″ in the pattern an arc is drawn through point 7″. Next, the true distance 3,7 is taken direct from the plan, and from point 3″ in the pattern an arc is described cutting the previous arc in point 7″.

The remainder of the pattern should readily be followed from this point, since the directions are similar throughout. The chief points to observe are that the true length diagonals from the base line must be taken down to the points on the vertical height line corresponding to the relative points on the line 2″10″ in the elevation, and that the true spacings for the bottom circle must be taken from the projected semicircle on the line 2′10′.

The additional back and side plates for the hopper are then marked off on the sides 3″7″, 7″11″ and 11″15″ in the pattern, as shown in Fig. 236.

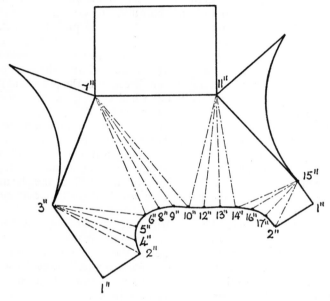

FIG. 236

Extraction Ducts

The arrangement of extraction ducts shown in Fig. 237 includes a few problems of pattern development which contain points of special interest. The branch piece shown at the top centre of the pipe system probably represents the simplest form of branch-piece design from the standpoint of practical construction. The pattern is easy to develop, and very little forming and shaping is required.

THE DEVELOPMENT OF THE BRANCH PIECE

The development of the pattern for the branch piece is shown in Fig. 238. It will be observed that the portion above the triangle in

Duct Layout with Developments

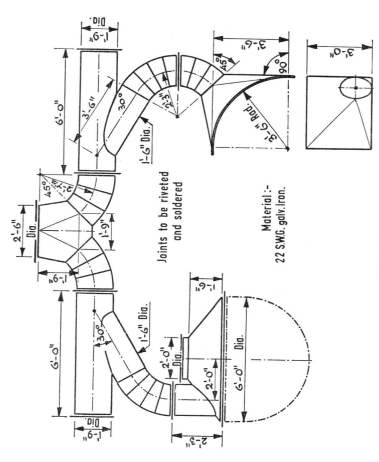

FIG. 237. *Extraction ducts*

the elevation, from a to d, is a portion of an oblique cylinder in which the true form of the cross-section at $abcd$ is elliptical, as shown at $a'b'c'd'$.

Prior to developing the pattern, describe a semicircle on the top edge as at $1,e$ in the elevation, divide it into six equal parts and project

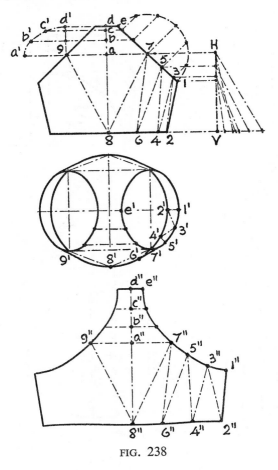

FIG. 238

the points perpendicularly back to the diameter $1,e$. Next project the true shape of the cross-section at $a'b'c'd'$, in which the horizontal distances of the points from the vertical centreline $d'9$ of the ellipse are equal to the corresponding widths of the semicircle. A vertical height line VH is erected, and the points 1, 3, 5 and 7 are projected

Duct Layout with Developments

horizontally to it. Assuming that the plan is satisfactorily obtained from the elevation, the surface of the lower curved portion is divided into triangles and the points numbered from 1 to 8 as shown in the Figure.

To develop the pattern, the true distance 1,2 is taken direct from the elevation, and marked off as from $1''$ to $2''$ in any convenient position to begin the pattern. Next, the plan length $2'3'$ is taken and marked off along the base line at right angles to VH. The true length diagonal is taken up to the point level with point 3 in the elevation, and from the point $2''$ in the pattern an arc is drawn through point $3''$. Now, one of the equal spacings is taken from the semicircle in the elevation, and from point $1''$ in the pattern an arc is described cutting the previous arc in point $3''$.

For the second triangle, the plan length $3'4'$ is taken and marked off along the base line from V. The true length diagonal is taken up to the point level with 3, and from point $3''$ in the pattern an arc is drawn through point $4''$. Next, the true spacing $2'4'$ is taken direct from the base circle in the plan, and from point $2''$ in the pattern an arc is described cutting the previous arc in point $4''$.

For the third triangle, the plan length 4,5 is taken and marked off along the base line from V. The true length diagonal is taken up to the point level with 5 in the elevation, and from point $4''$ in the pattern an arc is drawn through point $5''$. The next equal division is now taken from the semicircle in the elevation, and from point $3''$ in the pattern an arc is described cutting the previous arc in point $5''$.

The next three triangles, involving the diagonals 5,6, 6,7 and 7,8 should readily be obtained by following the above directions as they would apply to triangles up to the diagonal 7,8. In the next triangle, 7,8,9, it will be seen that the true length of the diagonal 8,9 will be the same as that of 7,8. Therefore in the pattern, the length of $7''8''$ is taken and from point $8''$ an arc is drawn through point $9''$. Next the true distance 7,9 is taken direct from the elevation (or, if preferred, the true distance $7'9'$ may be taken direct from the plan, since these lines are of the same true length), and from point $7''$ in the pattern an arc is described cutting the previous arc in point $9''$. The portion of the pattern beyond the line $8''9''$ will be a symmetrical repetition of that leading up to the line $7''8''$.

The portion of the pattern above line $7''9''$ will be most conveniently obtained by the parallel line method, in which the spacings $a''-b''-c''-d''$ are made equal to the spacings $a'-b'-c'-d'$ taken from the projected true shape of the cross-section in the elevation. Through points b'', c'' and d'', lines are drawn parallel to $7''9''$ and their true lengths marked off equal to the lines in the elevation which pass through the corresponding points b, c and d. Curves

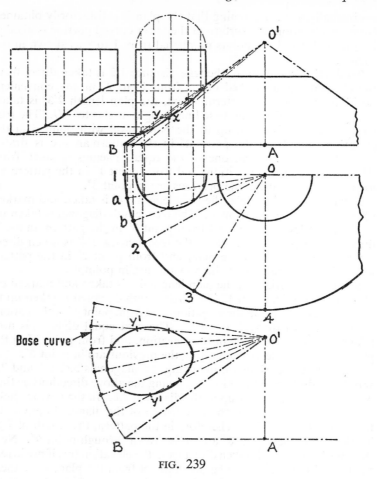

FIG. 239

drawn through the points obtained complete the pattern as shown in Fig. 238.

THE RIGHT CONICAL COVER

The intersection of the vertical cylindrical pipe and the right conical cover shown on the left-hand side of the elevation at Fig. 237 presents one or two interesting points in the method of determining the joint line. The joint line may be obtained by the method of radial lines on the cone, or, alternatively, by the method of cutting planes.

Duct Layout with Developments

The former method is shown in Fig. 239, and may be of particular interest from the fact that the plan of the cylinder falls entirely inside the radial line $O,2$, which forms the first division of the conic plan. Normally, for the purpose of pattern drafting, the quadrant $1,O,4$ would be divided equally into three parts, as at $1,2,3,4$. However, in the present case the semicircle representing half of the plan of the cylinder is not large enough to cut the first line $O,2$. In all such cases, the obvious solution is to further subdivide the first division into any convenient number of equal parts.

In the example shown in Fig. 239, the first division, 1 to 2, is subdivided into a further three equal parts, as at $1-a-b-2$. Then, as will be seen, the two extra lines aO and bO both cut the semicircle representing half of the plan of the cylinder. To continue with the construction, the points a, b and 2 are now projected vertically upwards to the base of the cone in the elevation, and from the points on the base of the cone, lines are drawn to the apex O'. Next, the points in the plan where the lines aO and bO cross the semicircle, are projected vertically upwards to the corresponding radial lines on the cone in the elevation. In this case there are two points on each line, four in all, which are projected upwards. Thus sufficient points are obtained in the elevation through which the intersection curve can be drawn.

Incidentally, it sometimes happens that a problem of this kind occurs in which the apex of the cone is too far off to be conveniently used as a point to which the radial lines might be converged. The correct positions of the radial lines may then be determined by dividing the top edge and the bottom edge into a corresponding number of equal parts and joining them as shown in Fig. 240. The procedure for determining the joint line is then exactly the same as in the example shown in Fig. 239.

The development of the pattern for the cylindrical connexion should readily be followed from the illustration, since once the joint line has been determined, the method of drafting the pattern is similar to that of a simple cylinder cut off at an angle.

The method of developing the true shape of the hole in the cone, however, depends on the method adopted for the determination of the joint line. In this example, use is made of the radial lines set out on the surface of the cone. The development of the hole is shown below the plan in Fig. 239. On either side of the centreline of the hole, spacings are set off along the base curve equal to the spacings $1,a$ and ab taken from the plan. In the next step, true distances are set off from the apex O', which mark the points where the radial lines are cut by the joint curve. These true lengths are obtained from the elevation by taking the distances from the apex O' down the outside of the cone to the points horizontally level with the corresponding points on the joint curve.

For example, in the elevation at Fig. 239, the true distance of the point x on the joint line from the apex O' is obtained by projecting point x horizontally to point y, which is on the outside line of the cone. The distance $O'y$ is then the true distance of point x from the

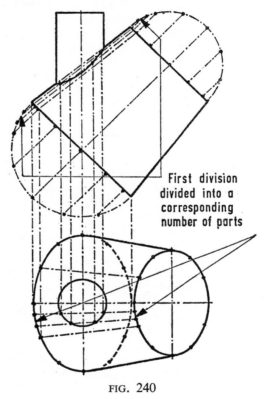

First division divided into a corresponding number of parts

FIG. 240

apex O', and is therefore marked off in the pattern on the corresponding radial line, as shown at $O'y'$.

Alternative Method by Cutting Planes The alternative method of obtaining the line of intersection, that of the method of cutting planes, is shown in Fig. 241. Any number of horizontal cutting planes may be taken at any convenient positions between the extremities of the joint line. In the present example four cutting planes are taken at a, b, c and d, as shown in the elevation. The

Duct Layout with Developments 309

plan of the cone cut through at either of the cutting planes is a circle, and in the bottom left-hand quadrant in Fig. 241 the arcs at a', b', c' and d' represent the plans of the cutting planes at a, b, c and d.

The plan of the cylinder at either of the cutting planes is a circle of the same diameter and in the same position. Hence, the smaller semicircle in the plan represents the half plan of the cylinder at any

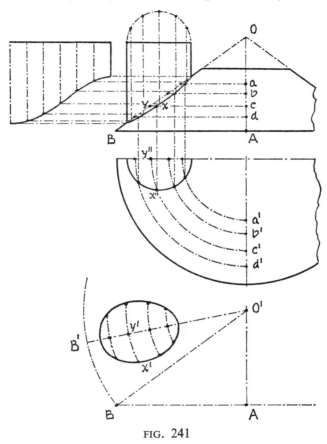

FIG. 241

cutting plane. From these conditions it will be seen that where the arcs a', b', c' and d' cut the smaller semicircle in the plan, the points represent positions of intersection on the various cutting planes, and may be projected vertically upwards to the respective cutting planes at a, b, c and d to obtain points through which the intersection line is drawn in the elevation.

The development of the pattern for the cylindrical connexion is exactly the same as that illustrated in Fig. 239, and should readily be followed from the diagram. The shape of the hole in the cone is shown below the plan in Fig. 241, in which the line $O'B'$ represents the line through the centre of the hole, and is equal to OB in the elevation. The radii of the arcs drawn across the hole are equal to the corresponding distances down the slant of the cone from the apex O to the position of the cutting planes in the elevation.

For example, the position of the cutting plane c occurs on the slant of the cone at y. Therefore the distance Oy is used as a radius to describe the arc through point y' in the contour diagram below the plan. The length of the arc on either side of point y' in the contour diagram, as at $y'x'$, should be equal to the length of the corresponding arc in the plan, as at $y''x''$. The lengths of the four arcs in the contour diagram, obtained in this way, should give sufficient points through which the contour of the hole may be drawn.

THE TRANSFORMER CONNEXION

The pattern for the transformer connexion on the right-hand side of Fig. 237 is shown developed in Figs. 242 and 243. In this problem the method of dividing the surface into triangles is of importance in ensuring a correct development.

For example, in the plan at Fig. 242 it will be seen that the part of the transforming piece between points 1, 3, 7 and 10 is divided up after the manner of an ordinary square-to-circle transformer, inasmuch as the corner points 3 and 7 are joined radially to the corresponding quadrants in the circle at the top, from points 1 to 6 and 6 to 10 respectively. The portion of the transforming piece between points 3, 6 and 7 lies in a vertical plane, and is therefore a flat section of the pattern. Flat triangles also occur between points 1, 3 and 22, and between points 7, 10 and 11. The remaining portion of the transforming piece between points 1, 10, 11 and 22 is a curved surface which is triangulated by the zigzag line between the two curved edges.

To develop the pattern, the true vertical height of the line 1 to 2 is taken from the elevation and marked off from $1''$ to $2''$ in any convenient position to begin the pattern. Next, the plan length 2,3 is taken and marked off at right angles to the vertical height VH. The true length diagonal is taken up to the point level with $2'$, and from point $2''$ in the pattern an arc is drawn through point $3''$. The true distance 1,3 is now taken direct from the plan, and from point $1''$ in the pattern an arc is described cutting the previous arc in point $3''$.

Duct Layout with Developments

For the next triangle, the plan length 3,4 is taken and marked off at right angles to *VH*. The true length diagonal is taken up to the point level with 4′, and from point 3″ in the pattern an arc is drawn through point 4″. Next, for the true distance between points 2 and 4, one of the equal spacings is taken from the semicircle in the

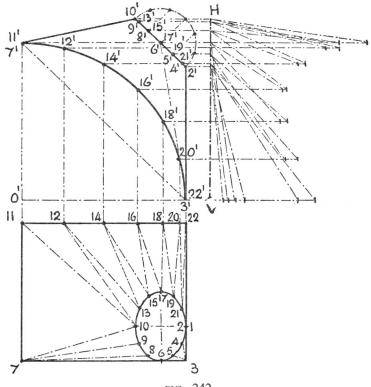

FIG. 242

elevation, and from point 2″ in the pattern an arc is described cutting the previous arc in point 4″.

For the third triangle, the plan length 3,5 is taken and marked off at right angles to *VH*. The true length diagonal is taken up to the point level with 5′, and from point 3″ in the pattern an arc is drawn through point 5″. Next, for the true distance between 4 and 5, one of the equal spacings is taken from the semicircle in the elevation, and from point 4″ in the pattern an arc is described cutting the previous arc in point 5″.

For the next triangle, the plan length 3,6 is taken and marked off at right angles to *VH*. The true length diagonal is taken up to the point level with 6′, and from point 3″ in the pattern an arc is drawn through point 6″. Then, one of the true spacings is taken from the semicircle in the elevation, and from point 5″ in the pattern an arc is described cutting the previous arc in point 6″.

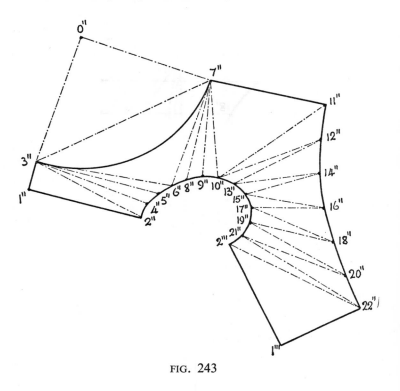

FIG. 243

The next triangle is an important step in the development. It will be seen in the elevation that the edge of the transforming piece between points 3′ and 7′ is a quarter circle with its centre of radius at 0′. However, since the portion between points 3′, 6′ and 7′ is flat, it will be more convenient to take a straight line from 3′ to 7′ and insert the quadrant in the pattern afterwards. It will also be seen that since the triangle 3′6′7′ is in a vertical plane, its true size and shape is represented in the elevation. Therefore, to proceed with the pattern, the true distance 6′7′ is taken direct from the elevation, and from point 6″ in the pattern an arc is drawn through

Duct Layout with Developments 313

point 7″. Next, the true distance 3′7′ is taken direct from the elevation, and from point 3″ in the pattern an arc is described cutting the previous arc in point 7″.

The quadrant in the pattern may now be drawn. To do this, the radius 0′3′ is taken direct from the elevation, and from point 3″ in the pattern an arc is drawn through point 0″. With the same radius, from point 7″ in the pattern, an arc is described cutting the previous arc in point 0″. Point 0″ may now be used as a centre for describing the quadrant in the pattern.

The development of the rest of the pattern is straightforward. The chief points on which to exercise care are the vertical heights of the diagonal lines, and the true spacings around the curved edges. For example, all the plan lines 7,8, 7,9, 7,10 and 10,11 are marked off at right angles to VH along the base line level with point 7′, and the true length diagonals taken respectively up to the points level with 8′, 9′ and 10′. Also, in triangulating the zigzag it must be observed that the base lines for the vertical heights will differ as the diagonals come down the curve from point 11′ to 22′. Thus, taking the diagonal 14,15 as an example, the plan length 14,15 is taken and marked off at right angles to VH along the base line level with point 14′. The true length diagonal is then taken up to the point level with 15′. The true spacings for the curve 11′ to 22′ are taken direct from the elevation, and those for the top edge 10′ to 21′ are taken from the semicircle in the elevation.

A Three-branch Duct System

The development of complex patterns does not necessarily refer to objects which are geometrically difficult to develop, but rather to objects which have complex surfaces. Such surfaces are not simple conic frustums or portions of cylindrical or prismatic bodies, but may be combinations of both, or even curvatures which cannot be assigned to either.

THE THREE-WAY JUNCTION PIECE

In the arrangement drawing, Fig. 244, a three-way branch piece is shown which is particularly suitable for branching to three points in close proximity. The branch piece divides directly from one circle to three circles, and may be made in one piece of metal, or preferably in two halves. The development of the pattern for this branch piece is shown in Fig. 245, where it will be seen that the half below the horizontal centreline in the plan, between points 1 and 29, will be similar to that above the centreline. Joints would therefore be made between points 1 and 2 at one end, and points 28 and 29 at the other.

In the practical forming of the branch piece, apart from the bends made in the folding machine along lines 8,11, 8,12, 16,19 and

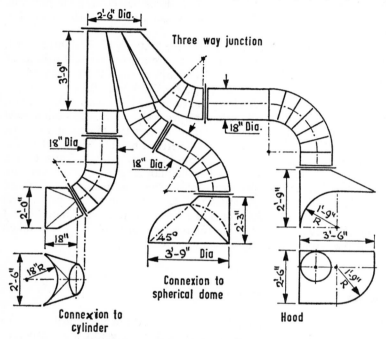

FIG. 244. *A three-branch duct system*

16,20, the work is simply that of shaping the metal by hand on a suitable bench bar.

Prior to developing the pattern, the surface of the branch piece is divided into triangles as shown in the plan and elevation, Fig. 245. The method of dividing the surface into triangles is important, as all lines must lie exactly along the surface of the geometrical body, or as nearly so as possible. Thus all the lines from 1 to 7 lie close to the theoretical surface, and all the lines radiating from point 8 to the top of the branch piece between points 7 and 15 lie exactly on the theoretical surface. Similarly, all the points radiating from point 16 to those between 15 and 23 lie exactly on the surface, while the remainder from 23 to 29 lie close to the surface.

Duct Layout with Developments 315

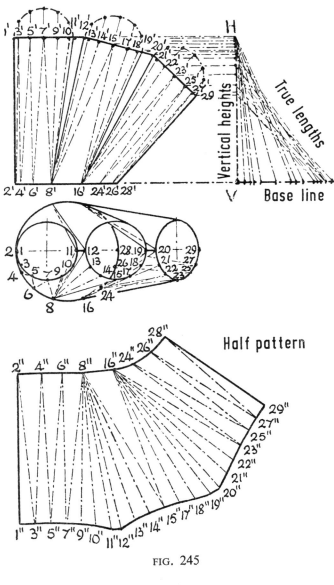

FIG. 245

To begin the pattern, the true distance 1′2′ is taken direct from the elevation and marked off in any convenient position, as from 1″ to 2″. Next, the plan distance 2,3 is taken and marked off along the base line at right angles to the vertical height VH. The true length diagonal is taken up to the point level with 3′, and from point 2″ in the pattern an arc is drawn through point 3″. Now, one of the true spacings is taken from the semicircle above 1′10′, and from point 1″ in the pattern an arc is described cutting the previous arc in point 3″.

For the second triangle, the plan length 3,4 is taken and marked off along the base line at right angles to the vertical height VH. The true length diagonal is taken up to the point level with 3′, and from point 3″ in the pattern an arc is drawn through point 4″. Next, the true spacing 2,4 is taken direct from the plan, and from point 2″ in the pattern an arc is described cutting the previous arc in point 4″.

For the third triangle, the plan length 4,5 is taken and marked off along the base line at right angles to the vertical height VH. The true length diagonal is taken up to the point level with 5′, and from point 4″ in the pattern an arc is drawn through point 5″. Next, one of the equal spacings is taken from the semicircle above 1′10′ in the elevation, and from point 3″ in the pattern an arc is described cutting the previous arc in point 5″.

The remainder of the pattern is straightforward triangulation, and it is unnecessary to give a full description since it would involve a repetition of terms for each triangle. To complete the pattern, take in turn each plan length which passes from the large base circle to one of the smaller top circles, and mark it off at right angles to the vertical height. In taking the true length diagonals, care must be exercised in taking them up to the correct points on the vertical height line corresponding to the heights of the relative points on the top circles. For example, take the line 16,22 and mark it off along the base line at right angles to VH. The true length diagonal is then taken up to the point level with 22′ in the elevation, and from point 16″ in the pattern an arc is drawn through point 22″. Next, one of the true spacings is taken from the semicircle above 20′29′, and from point 21″ in the pattern an arc is described cutting the previous arc in point 22″.

CONNEXION TO CYLINDER

The connecting piece to a vertical cylindrical body shown in Fig. 244 is of interest from the standpoint of pattern development on account of the difference in the method of dividing the top and

Duct Layout with Developments

bottom portions into triangles. The development is shown in Fig. 246. It will be noted that in the plan the upper half of the circular top, at 7,12,15, is connected to the curved arc of the cylinder at 9,14,16, and that the lower half of the circular top, at 7,1,15, is connected to the curved arc of the cylinder lower down at 2,4,6,8.

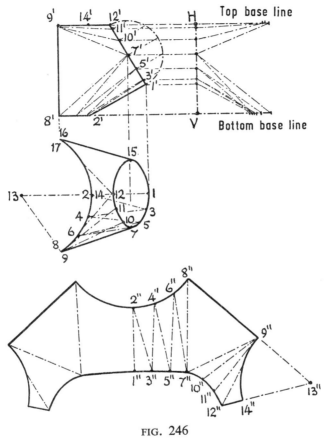

FIG. 246

Since, at the top, the point 12 lies in the same horizontal plane as the curve of connexion at 9,14,16 the figure lying between the points 12 and 9, 14, 16 will be flat. Also, the triangle in the vertical plane at 7,8,9 will be flat, as also will be the triangle on the opposite side at 15, 16, 17. Then, it should be observed, the portion lying between the points 12, 9 and 7 will be conical, and the intermediate lines 9,10 and 9,11 will lie exactly on the surface. On the bottom portion of the connecting piece no point on the circular end between points 1 and 7 lies in the same plane as the curve of connexion from point 2 to 8. Therefore the surface between the two curves is triangulated by the zigzag line 1,2,3,4,5,6,7,8. The joint is arranged to occur on the top between points 12 and 14, which makes the pattern symmetrical about the line 1,2.

To develop the pattern, the true length distance 1'2' is taken direct from the elevation and marked off as from 1″ to 2″ in any convenient position. Next, the plan length 2,3 is taken and marked off along the bottom base line at right angles to VH. The true length diagonal is taken up to the point level with 3', and from point 2″ in the pattern an arc is drawn through point 3″. Now, one of the true spacings is taken from the semicircle on 1'12', and from point 1″ in the pattern an arc is described cutting the previous arc in point 3″.

For the next triangle, the plan length 3,4 is taken and marked off along the base line at right angles to VH. The true length diagonal is taken up to the point level with 3', and from point 3″ in the pattern an arc is drawn through point 4″. Next, the true distance 2,4 is taken direct from the plan and from point 2″ in the pattern an arc is described cutting the previous arc in point 4″.

For the third triangle, the plan length 4,5 is taken and marked off at right angles to VH. The true length diagonal is taken up to the point level with 5' and from point 4″ in the pattern an arc is drawn through point 5″. Now, one of the true spacings is taken from the semicircle on 1'12', and from point 3″ in the patternan arc is described cutting the previous arc in point 5″.

The instructions for the development of the next three triangles as far as line 7″8″ are a repetition of the first three except that the appropriate numbers for the points and spacings should be used.

The next triangle 7″8″9″ is obtained by taking the plan length 7,9 and marking it off along the top base line. The true length diagonal is taken, this time down to the point level with 7', and from point 7″ in the pattern an arc is drawn through point 9″. Next, the true distance 8'9' is taken direct from the elevation and from point 8″ in the pattern an arc is described cutting the previous arc in point 9″.

The development of the next three triangles as far as line 9″12″ should readily be followed from this point. The chief condition to be observed is that the plan lengths of lines 9,10, 9,11 and 9,12 should be marked off along the top base line, and the respective true diagonals taken downwards to the corresponding points on VH.

Finally, the flat portion between points 9″, 12″ and 14″ is obtained by taking the true distance 9,13 direct from the plan, which is the radius of the curve 9,14, and from point 9″ in the pattern an arc

Duct Layout with Developments

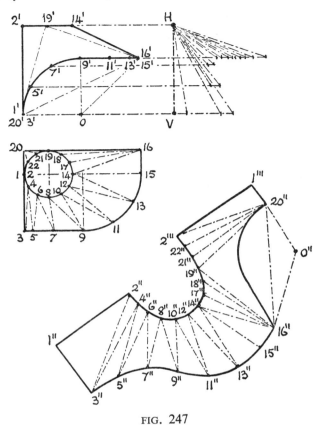

FIG. 247

is then drawn through point 13″. Next, the true distance 12,13 is taken direct from the plan, and from point 12″ in the pattern an arc is described cutting the previous arc in point 13″. Now, with point 13″ as centre, the arc 9″14″ is drawn, which completes the half pattern.

THE HOOD

The hood shown on the right in Fig. 244 represents a useful type which fits into a corner position or flat against a wall. In the development of the pattern, which is shown in Fig. 247, the method of obtaining the curved edge at the back between points 16″ and 20″

is worthy of attention, as is also the method of dividing the surface of the hood into triangles.

It will be noted that from points 16 and 20 the triangles radiate to the corresponding quarter circles in the top between points 14 and 2. The portion between points 14, 16 and 19 is conical in form and the lines 16,14, 16,17, 16,18 and 16,19 lie exactly on the surface. Similarly, the portion between points 19, 20 and 2 is conical in form and the lines 20,19, 20,21, 20,22 and 20,2 lie exactly on the surface.

On the other side of the hood, the zigzag line oscillates between the two curved edges 2,8,14 and 3,9,15. In this case it is not possible for all the lines to lie on the surface. The portion of the surface between lines 2,3 and 8,9 is of complex curvature between the two quarter circles 2,8 and 3,9, which lie in planes at right angles to each other. The next section of the surface between lines 8,9 and 14,15 is a portion of an oblique conic frustum since the two quarter circles 8,14 and 9,15 lie in planes parallel to each other. In consequence, the lines 8,9, 10,11, 12,13 and 14,15 lie exactly on the surface, while the diagonals 9,10, 11,12 and 13,14 lie close to the surface.

The above analysis of the surface of the hood is a necessary preliminary to the correct method of dividing it into triangles.

To develop the pattern, the true distance 1′2′ is taken direct from the elevation and marked off from 1″ to 2″ in any convenient position to begin the pattern. Next, the plan length 2,3 is taken and marked off along the bottom base line at right angles to *VH*. The true length diagonal is taken up to the top point *H*, and from point 2″ in the pattern an arc is drawn through point 3″. Now, the true distance 1,3 is taken direct from the plan, and from point 1″ in the pattern an arc is described cutting the precious arc in point 3″.

For the second triangle, the plan length 3,4 is taken and marked off along the bottom base line at right angles to *VH*. The true length diagonal is taken up to the top point *H*, and from point 3″ in the pattern an arc is drawn through point 4″. Next, the true spacing 2,4 is taken direct from the plan, and from point 2″ in the pattern an arc is described cutting the previous arc in point 4″.

For the third triangle, the plan length 4,5 is taken and marked off this time from *VH* along the base line level with point 5′. The true length diagonal is taken up to the top point *H*, and from point 4″ in the pattern an arc is drawn through point 5″. Next, the true spacing 3′5′ is taken direct from the elevation and from point 3″ in the pattern an arc is described cutting the previous arc in point 5″.

From this point the development of the pattern is straightforward triangulation as far as line 16″19″. The chief points to observe are that true spacings at the top are taken from the circle in the plan between points 2 and 19; that true spacings at the bottom are taken

Duct Layout with Developments 321

from the quarter circle 3' to 9' in the elevation, the quarter circle 9,15 in the plan and the space 15,16 in the plan; and that the appropriate base lines must be used, level with points 7', 9', and so on.

To continue the pattern from the line 16"19", the plan length 19,20 is taken and marked off along the bottom base line at right angles to VH. The true length diagonal is taken up to the top point H, and from the point 19" in the pattern an arc is drawn through point 20". The next step is an important one. The triangle between the points 16", 19" and 20" contains the curved edge which corresponds to that in the elevation from 16' to 20'. As a preliminary to obtaining this curve, the straight line distance from 16' to 20' is taken direct from the elevation and from point 16" in the pattern an arc is described cutting the previous arc in point 20".

Now, to obtain the curve, the true distance from 20' to 0 is taken direct from the elevation, and from point 20" in the pattern an arc is drawn through point 0". Next, the true distance 16'0 is taken direct from the elevation, and from point 16" in the pattern an arc is described cutting the previous arc in point 0". Point 0" is now the centre from which the arc from point 20" is described. This arc meets the straight line tangentially drawn from point 16". The remainder of the pattern should readily be followed from this point.

CONNEXION TO SPHERICAL DOME

The oblique conical connexion piece to the spherical dome shown in the middle of Fig. 125 (p. 135) has one or two points of special interest related as much to solid geometry as to surface development.

Normally the base of an oblique cone is regarded as a circle, while the central axis of the cone leans at an angle inclined to the plane of the base. In the example shown in Fig. 244, the back of the cone is perpendicular to the base, while the central axis leans from the centre of the base to the centre of the circular top. The development of the pattern is shown in Fig. 248, wherein the elevation of the conical connexion will be seen at $BDEF$, the apex of the cone at A, and the projected base of the cone at BC. The base at BC also represents the base of the spherical dome BDC.

In any oblique cone there are two angular positions in which a circular cross-section occurs. The base at BC represents one position, since that is already a circle by arrangement. The other position, known as the subcontrary section, is such that the angle ABC is formed on the opposite side of the cone, as at ADB. It will be noted that the curve BDC is a semicircle in which the point D is on the circumference. In consequence the angled formed by the

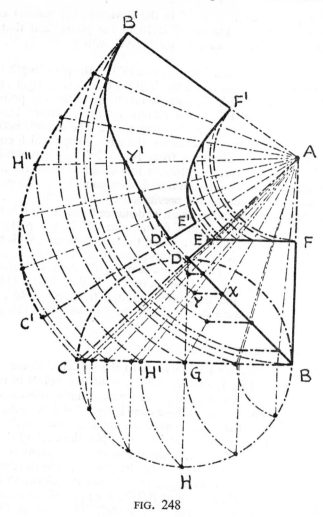

FIG. 248

chords *BD* and *DC* is a right angle. It follows from this that, as *ADC* is a straight line and *BDC* is a right angle, the angle *ADB* is also a right angle. But angle *ABC* is also a right angle. From these conditions, it follows that the cross-section of the cone at *BD* is the subcontrary section and is therefore a circle. Further, since the cross-section of the cone at *BD* is a circle, it exactly fits the cross-section of the sphere at *BD*, which is also a circle.

Duct Layout with Developments

From these considerations it may be seen that the pattern for the oblique conical connexion may be developed in either of two ways, shown respectively in Figs. 248 and 249. The more usual method is the one shown in Fig. 248 in which BC is regarded as the base.

To develop the pattern, the semicircle BHC is described on the base BC to represent half of the plan of the cone. As AB is perpendicular to BC, the apex A falls on B in the plan. The semicircle BHC is divided into the usual six equal parts, and from point B arcs are described from the points on the semicircle to the base line BC. Then with apex A as centre, arcs are drawn from the points on the base BC into the pattern. Next, one of the equal divisions is taken from the semicircle and, beginning at any point C' on the outside arc from point C, the divisions are stepped over from one arc to the next until the inside arc from point B is reached. Through these points the base curve $C'H''B'$ is drawn, and a line is drawn from each point to the apex A.

To obtain the top curve $E'F'$, lines are drawn from the points on the base BC to the apex A, and from the points where these lines cross the top edge EF, arcs are swung into the pattern, using the apex A as centre, to meet the corresponding lines in the pattern from the base curve $C'H''B'$. Points are thereby obtained through which the top curve $E'F'$ is drawn in the pattern.

It now remains to determine the curve $D'B'$ in the pattern. Bearing in mind that all the points on the curve $D'B'$ must be at true distances from the apex A, and that the radial lines in the pattern are already true length lines, the next step is to determine the true distances of the required points from the apex in the elevation. To do this, lines are drawn from all the points on the semicircle BHC perpendicularly to the base BC of the cone, and from the points thus obtained on the base, lines are drawn to the apex A. These lines are "elevation" lines, and their corresponding "true length" lines are those drawn to the apex A from the points swung round to the base line from the centrepoint B. For example, from the point H, the line HG is drawn perpendicularly to the base BC. Then the line GA is the elevation line. The corresponding true length line is shown at $H'A$, which is obtained by describing the arc HH' and then drawing $H'A$ to the apex A. Now, the elevation line GA crosses the joint line BD in point x, and the true distance of the point x from the apex A is obtained by drawing xy horizontally to meet the corresponding true length line in point y. Then Ay is the true distance of x from the apex A. Now, from the apex A, the distance Ay is swung into the pattern to meet the true length line $H''A$ in the point y'.

This process repeated from the other points where the elevation lines cross BD should give the required points on the curve $D'B'$ in the pattern. Then the figure $D'B'F'E'$ represents half of the pattern for the connecting piece to the spherical dome.

The Alternative Method Since the joint line at BD between the hemisphere and the connecting piece is shown to be a circle, it

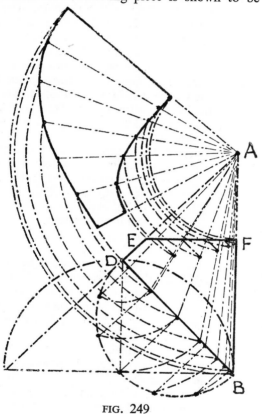

FIG. 249

would be correct and in order to treat the portion DAB as an oblique cone on the base DB with an inclined cut-off at EF.

This method of treatment is shown in Fig. 249. A description of the process of development would be similar to that of Fig. 248, except that in this case the base line BD is now regarded as horizontal. Care is also needed to ensure that lines are drawn parallel to the base BD from the correct points where the elevation lines cut the

Duct Layout with Developments

line *EF* to meet the corresponding true length lines before swinging them into the pattern.

Exhaust Ducts

The section of ductwork shown in Fig. 250 comprises three problems of pattern development, the junction piece at the top, the intersection of the vertical pipe with the hemispherical dome shown

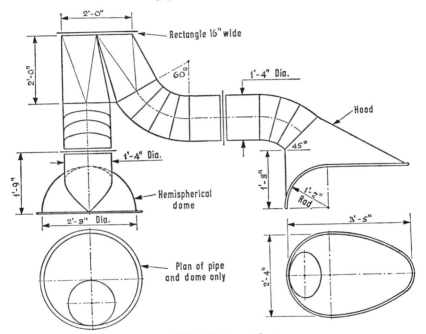

FIG. 250. *Exhaust ducts*

below, and the hood at the end on the right-hand side. The development of the pattern for the junction piece is similar to that shown in Fig. 144 (p. 163). The only differences are that in the present case the base, inverted, is a rectangle instead of a square, and the diameters of the branches are equal instead of unequal.

PIPE CONNEXION TO DOME

The determination of the joint line and the development of the pattern for the pipe intersecting the dome are shown in Fig. 251.

The elevation in Fig. 251 is equivalent to a side view, to facilitate the determination of the intersection line and the unrolling of the pattern for the pipe. The intersection line is obtained by the method of cutting planes. The quadrant on the hemisphere is divided into six equal parts, as from point 1 to 7, and horizontal lines are drawn from these points to represent the cutting planes through the dome and the cylinder.

Next, the points from 1 to 7 are projected vertically to the horizontal centreline in the plan. Then, from the centre of the circle

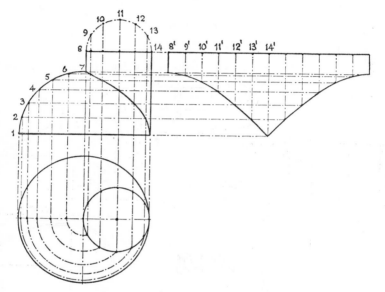

FIG. 251. *Development of pattern for pipe connexion to dome*

which represents the dome, arcs are drawn to meet the smaller circle which represents the plan of the cylinder. From the points thus obtained on the smaller circle, lines are drawn vertically upwards to meet the corresponding horizontal lines which represent the cutting planes. Points are thereby afforded through which the intersection curve is drawn in the elevation.

Now, on the top end of the cylinder in the elevation, a semicircle is drawn and divided into the usual six equal parts. From the points on the semicircle, lines are drawn vertically downwards parallel to the centreline of the cylinder to meet the intersection curve. It should be noted that in this case the vertical lines on the

Duct Layout with Developments

cylinder meet the horizontal planes at the same points as those obtained by the lines drawn vertically upwards from the smaller circle in the plan. This is because the quadrant from 1 to 7 is divided into six equal parts to use as points for the cutting planes. Had the quadrant been divided differently for the cutting planes, as would have served equally well, then the vertical lines drawn from the top of the cylinder would not have coincided with those from the circle in the plan.

The pattern is now unrolled from the cylinder in the elevation, and is of straightforward parallel line development.

THE HOOD

The development of the hood is shown in Fig. 252, in which the pattern is obtained by fairly straightforward triangulation. The joint is arranged to occur on the short side at the vertical end of the hood, and as the pattern will be symmetrical about the horizontal centreline, the numbering for the development begins from 1 to 2 on the opposite side of the hood. Both halves of the pattern may then be developed simultaneously on each side of the centreline 1″2″.

From the arrangement drawing in Fig. 250, it will be noted that the three-piece lobster-back bend joins the top of the hood, and that the shape of the top will be circular. Therefore, in Fig. 252, a semicircle is described on the top edge of the hood, divided into six equal parts, and the points projected back to the edge. The points on the edge are then dropped into the plan to obtain the ellipse numbered 2 to 14.

The bottom edge of the hood in the plan is egg-shaped, comprising half a circle on the left and half an oval on the right. One half of the plan is divided up for triangulation, the oval portion into three equal parts and the circular quadrant into three equal parts. The points, numbered from 1 to 13, are projected into the elevation to give points 1′ to 13′. The points on the top and bottom edges of the hood are now joined by the zigzag line 1,2,3,4, . . . 12,13,14, thereby dividing the surface into triangles.

The vertical height line VH is drawn on the left of the elevation, and all the points on the top and bottom edges are projected horizontally to VH. Those on the bottom edge are extended beyond VH to serve as base lines.

The development of the pattern is begun by taking the true distance 1′ to 2′ direct from the elevation and marking it off in any convenient position as from 1″ to 2″. The next distance from 2 to 3 is taken from the plan and marked off from VH along the horizontal line

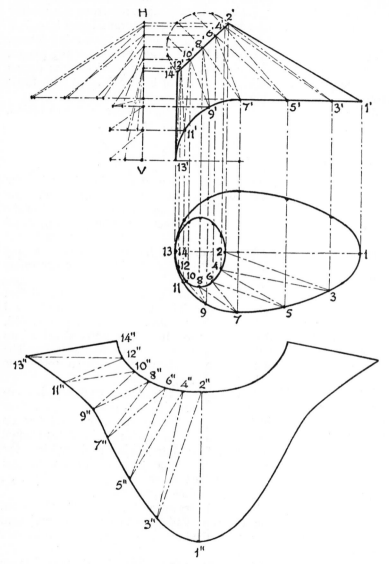

FIG. 252. *Development of pattern for hood*

Duct Layout with Developments

level with point 3′ in the elevation. The true length diagonal is taken up to the point level with 2′, and from point 2″ in the pattern an arc is drawn through point 3″. Next, the true spacing from 1 to 3 is taken direct from the plan and from point 1″ in the pattern an arc is described cutting the previous arc in point 3″. This completes the first triangle.

For the second triangle, the plan length 3,4 is taken and marked off from VH along the horizontal line level with 3′ in the elevation. The true length diagonal is taken up to the point level with 4′, and from point 3″ in the pattern an arc is drawn through point 4″. Next, the true spacing between points 2 and 4 is taken from the semicircle on the top edge in the elevation, and from point 2″ in the pattern an arc is described cutting the previous arc in point 4″. This completes the second triangle.

The next four triangles round the pattern between points 7″ and 8″ are obtained in the same way, taking the true spacings from the semicircle at the top in the elevation, and from the oval between points 3 and 7 in the plan.

For the next triangle, the plan length from 8 to 9 is taken and, this time, is marked off from VH along the horizontal line from point 9′ in the elevation. The true length diagonal is taken up to the point level with 8′, and from point 8″ in the pattern an arc is drawn through point 9″. Next, it must be noted that the plan distance between points 7 and 9 is not its true distance because these points lie on the curve 7′9′11′13′ in the elevation. Therefore to obtain the true distance between points 7 and 9, the plan length must be triangulated against the vertical height. Thus the plan distance 7 to 9 is taken and marked off from VH along the line level with 9′ in the elevation. The true length diagonal is taken up to the point level with 7′, and from point 7″ in the pattern an arc is described cutting the previous arc in point 9″.

The remaining triangles are obtained in a similar manner. The chief precaution to observe is to triangulate the plan distances 9,11 and 11,13 against their respective vertical heights in order to obtain their true lengths for the pattern.

Fume Ducts

The next arrangement of a section of ductwork is given in Fig. 253, and the items for development included in the system are a branch connexion, a transformer connexion to a cylindrical body, a two-way junction piece, and an oblique conical connexion to a right

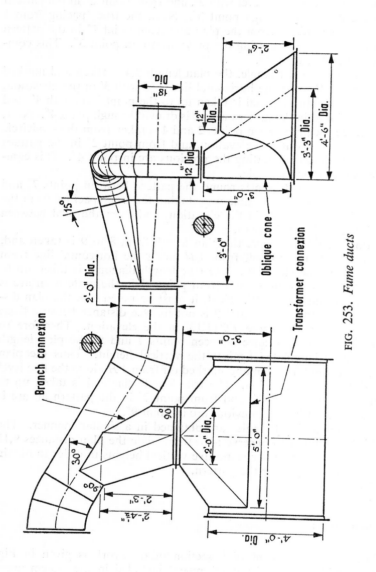

FIG. 253. *Fume ducts*

Duct Layout with Developments

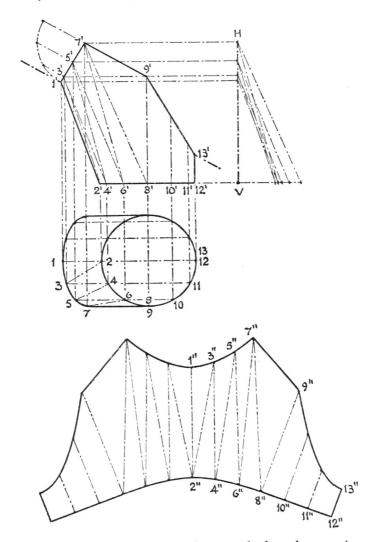

FIG. 254. *Development of pattern for branch connexion*

conical body. The development of the two-way junction piece is not given here as this type of branch piece is adequately dealt with elsewhere. The pattern for the branch connexion is shown developed in Fig. 254.

THE BRANCH CONNEXION

The composition of this branch piece needs to be carefully analysed before attempting the development of the pattern. For instance, the base of the branch piece is a circle as on the diameter 2 to 12 in the plan. From the half circle 8' to 12' in the elevation, that section of the branch piece rises vertically to meet the inclined main pipe at 30°. It is evident from these conditions that the portion 8'9'13'12' of the branch piece constitutes a part of a right cylinder and may therefore be developed by the parallel line method. Added to this, two flat triangles occur as between the points 7', 8' and 9' in the elevation. The remaining portion of the branch piece, as between the points 1', 2', 8' and 7', will need to be developed by triangulation. It might here be observed that had the line 7'8' been drawn parallel to 1'2' to meet the centreline of the main pipe a little lower down, and the top part of the joint line from 1' to 7' been adjusted to suit, that portion of the branch piece would have conformed to half an oblique cylinder. Since the joint line from 1' to 7' is at 90° to the axis of the main pipe its shape will be semicircular, and is dropped into the plan as a semi-ellipse as shown in the Figure.

Since the plan of the branch piece is symmetrical about the horizontal centreline, only the bottom half is divided up for development. The numbering of the points is begun at the back, or the long side, so that the two halves of the pattern may be developed simultaneously, thereby bringing the seam at the opposite end or on the short side.

To begin the pattern, the outside line from 1' to 2' is taken direct from the elevation and marked off in any convenient position, as at 1'' to 2''. Next, the plan length from 2 to 3 is taken and marked off along the base line at right angles to the vertical height VH. The true length diagonal is taken up to the point level with 3', and from point 2'' in the pattern an arc is drawn through point 3''. The true spacing in the pattern from 1'' to 3'' is obtained from the quadrant on the top edge in the elevation, and from point 1'' an arc is drawn through point 3''.

For the next triangle, the plan length from 3 to 4 is taken and marked off from V along the base line. The true length diagonal is taken up to the point level with 3', and from point 3'' in the pattern an arc is drawn through point 4''. Now, the true spacing is taken from 2 to 4 direct from the base circle in the plan, and from point 2'' in the pattern an arc is described cutting the previous arc in point 4''.

The remainder of the triangulation is similar, though care must

Duct Layout with Developments

be taken to triangulate the lines 4,5 and 6,7 up to the points on VH level with points 5′ and 7′ respectively. The section developed by triangulation is completed at line 7″8″ in the pattern.

The next triangle, 7″8″9″, is obtained direct from the elevation, for since the triangle lies in a vertical plane it is seen at its true size in the elevation. Therefore the distance 8′9′ is taken direct from the elevation and from point 8″ in the pattern an arc is drawn through point 9″. Then the distance 7′9′ is taken from the elevation and from point 7″ in the pattern an arc is drawn cutting the previous arc in point 9″.

The remainder of the pattern is developed by the parallel line method. From point 8″ in the pattern a line is drawn at right angles to 8″9″, and the spacings 8″–10″–11″–12″ are made equal to the distances 8,10, 10,11, and 11,12 in the plan. Lines are now drawn parallel to 8″9″ from the points 10″, 11″ and 12″ in the pattern. Next, the heights from points 10′, 11′ and 12′ up to the joint line 9′13′ are taken direct from the elevation, and marked off on the corresponding lines from points 10″, 11″ and 12″ in the pattern. Points are thereby afforded through which the joint curve from 9″ to 13″ is drawn in the pattern. This completes the pattern for the branch connexion.

THE TRANSFORMER CONNEXION

The setting out for the development of the transformer connexion needs a side elevation in order to determine the position of the bottom edge of the transformer. Thus half of a side elevation is shown on the right-hand side of Fig. 255. The position of point 1^0, which represents the bottom edge of the transformer, is determined by drawing the side $1^0 2^0$ of the transformer as a tangent to the semicircle which represents the body of the cylinder, and then, from the centre 0^0, drawing $0^0 1^0$ at right angles to $1^0 2^0$. The point 1^0 then represents the exact position where the side $1^0 2^0$ touches the cylinder.

The front elevation and plan are drawn and divided up as shown for the development of a half pattern with the seams on the horizontal centreline on each side of the transformer. The vertical height line VH is drawn on the right of the side elevation. Only one quarter of the transformer is numbered for development, as both quarters of the pattern are similar and may be developed simultaneously.

To begin the pattern, the plan length from 1 to 2 is taken and marked off along the base line level with 1^0 at right angles to VH. The true length diagonal is taken up to the top point H, and the distance 1″2″ marked off in any convenient position in the pattern. It should be evident from the drawing that the two lines 1,2 and 1,3

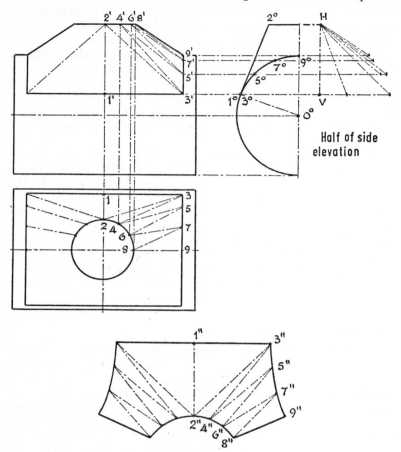

FIG. 255. *Development of half pattern for transformer connexion to cylindrical container*

form a right angle. Therefore, through point 1″ in the pattern, the line 1″3″ is drawn at right angles to 1″2″, and the distance 1″3″ made equal to 1,3 in the plan. The points 2″ and 3″ are now joined.

For the next triangle, the plan length 3,4 is taken and marked off along the line level with 3^0 at right angles to *VH*. The true length diagonal is taken up to the top point *H*, and from point 3″ in the pattern an arc is drawn through point 4″. Next, the true spacing from 2 to 4 is taken direct from the plan, and from point 2″ in the pattern an arc is described cutting the previous arc in point 4″.

For the third triangle, the plan length 4,5 is taken and this time marked off along the line level with 5^0 at right angles to VH. The true length diagonal is taken up to the top point H, and from point $4''$ in the pattern an arc is drawn through point $5''$. Next, the true spacing from 3^0 to 5^0 is taken direct from the side elevation, and from point $3''$ in the pattern an arc is described cutting the previous arc in point $5''$.

The remaining triangles to complete the pattern are obtained in a similar manner. The true spacings from $2''$ to $8''$ are all taken from the corresponding divisions in the plan, and the true spacings from $3''$ to $9''$ are all taken from the corresponding divisions on the quadrant in the side elevation. The diagonals $5''6''$, $6''7''$, $7''8''$ and $8''9''$ are each obtained by triangulating their plan lengths against their respective vertical heights.

THE OBLIQUE CONIC CONNEXION

The joint line between the oblique conic connexion and the right conic cover shown in Fig. 253 is most easily determined by the method of cutting planes, as shown in Fig. 256. Four cutting planes as at a, b, c and d, are marked off at convenient positions between the top and bottom extremities of the joint line, and are drawn horizontally or parallel to the base line of the right conic cover. The cutting planes are drawn through both cones as shown in the Figure. At each cutting plane, the two cones will present two circles in the plan, which intersect or cut each other at points corresponding to the position of the cutting plane.

To determine the circles in the plan, the points a, b, c and d are first dropped vertically to the horizontal centreline in the plan, and from the centre of the right cone in the plan, circles are drawn which represent the plan of the right cone at each of the cutting planes. Next, it should be observed that although the plan of the oblique cone at each cutting plane will be a circle, the centres of these circles will not occur at a single point as is the case with a right cone. Therefore to determine the centres of the circles for the oblique cone, the points where the cutting planes pass through its centreline in the elevation are dropped vertically to the horizontal centreline in the plan. The points thus obtained on the horizontal centreline are the centres from which the circles are drawn. From the points where these circles cut the corresponding circles on the right cone, lines are drawn vertically upwards to intercept the lines which represent the corresponding cutting planes in the elevation. Points are thereby afforded through which the joint curve in the elevation is drawn.

There are two methods by which the pattern for the oblique cone may be developed. One is to use the points already obtained on the joint line and draw radial lines through them to the apex of the cone. The other is to ignore the points on the joint line and develop the pattern by the usual method of oblique conic development.

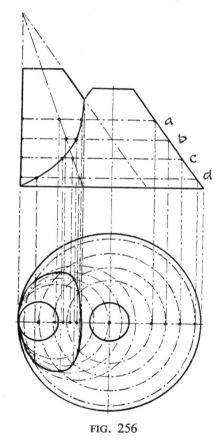

FIG. 256

The development shown in Fig. 257 is based on the first method mentioned above.

Having obtained the joint line as in Fig. 256, radial lines are drawn from the apex through the points on the joint line to the base of the cone. A semicircle is drawn on the base to represent a half plan, and from the points obtained on the base, lines are drawn perpendicularly downwards to the semicircle. It should now be

Duct Layout with Developments

specially noted that the spacings around the semicircle obtained by this method are not equal, and must necessarily depend on the positions of the points on the joint line through which the radial lines are drawn.

Next, from the apex A' in the plan the points obtained on the semicircle are swung round to the base, and true length lines are

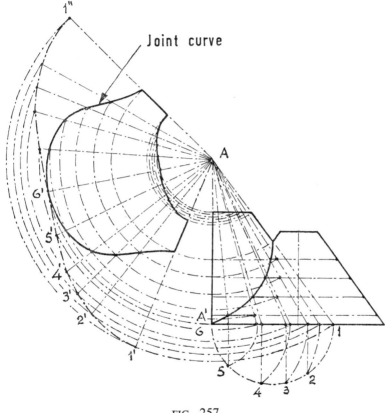

FIG. 257

drawn therefrom to the apex A. Then from the apex A the true length lines are swung into the pattern.

The next step should be carefully observed. In order to obtain the joint line on the short side at the top of the cone, the spacings are taken from the semicircle, as from 1 to 2, 2 to 3, 3 to 4, 4 to 5 and 5 to 6. Then, starting on the outside arc, as at 1' in the pattern,

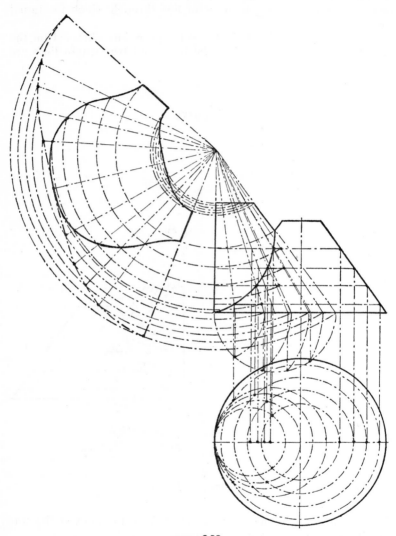

FIG. 258

Duct Layout with Developments

the spacings are stepped over from one line to the next until the inside arc is reached, and then the process repeated in the reverse order back to the outside arc. Lines are then drawn from these points to the apex A.

Now, to obtain the joint curve in the pattern, the true distances are first obtained in the elevation by projecting the points on the joint line horizontally to the corresponding true length lines. The points thus obtained are swung into the pattern to the corresponding true length lines, thereby obtaining the points through which the joint curve is drawn as shown in Fig. 257.

The top curve in the pattern is obtained by swinging the points on the true length lines from the top edge of the frustum in the elevation into the pattern to meet the corresponding true length lines. Points are thereby obtained through which the top curve is drawn.

In Figs. 256 and 257 the determination of the joint line by the method of cutting planes and the development of the pattern are shown independently, chiefly to make it easier to follow the directions given above. The two processes may, however, be combined in one illustration as shown in Fig. 258.

Section of Pipework with Branch at Complex Angle

The section of pipework shown in Fig. 259 includes a branch piece of the breeches type which fits into the system at a complex angle. The solution of this problem depends primarily on the setting out of the branch piece with its three-segment bend so that the pattern for the breeches piece can be developed and the true angles of the segments for the bend may be determined. The setting out of the branch piece is shown in Fig. 260 and the development of the patterns in Fig. 261.

The aim is to obtain a view of the branch piece and the attached lobster-back bend with their axes lying flat in the plane of the paper. In Fig. 260, note that the points on the centreline of a section of the side elevation are lettered m^0, n^0, o^0, p^0, q^0, and that these points bear similar lettering through the front elevation and the two projections.

The first projection is made in the direction of the arrow Y, which is at right angles to the base of the breeches piece at p. The points m, n, o, p and q are projected forward into the projection from the front elevation. The base at p then becomes a circle in the first projection and its centre is marked p'. A base line at right angles to the direction of projection is made through point p as shown in the Figure, and also a datum line through p' is made parallel to the base line through *p*. Next, a vertical datum line is drawn through p^0 in the side elevation, and the horizontal distances from the datum line to the centrepoints m^0, n^0 and o^0 are taken and marked forward on the corresponding projection lines from the datum line in the first projection. It should be noted that the position of point q^0 is taken from the datum line and marked backward from the datum line in the first projection.

Now, in the first projection, all these points should occur on a straight line from m' through p' to q'. Also, the circle with its centre

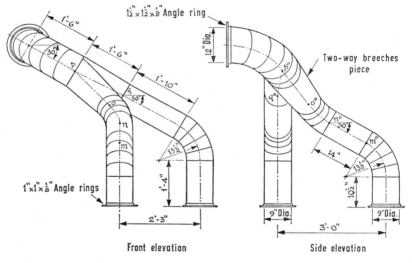

FIG. 259

at q', which represents the circular edge of the straight branch, should occur with its circumference touching the circumference of the larger circle at the end of the straight line through q'. This, then, produces a "straight-backed" breeches piece in the second projection.

The second projection is made in the direction of the arrow *X*, which is at right angles to the centreline $m'p'$. Therefore, from all the points m', n', o', p' and q', lines are drawn into the second projection, and a base line is drawn at right angles to the direction of projection, as shown in the Figure. Then from the front elevation, the distances from the base line through p to the points m, n, o and q are taken and marked off on the corresponding lines from the base line in the second projection to obtain the points m'', n'', o''

Duct Layout with Developments

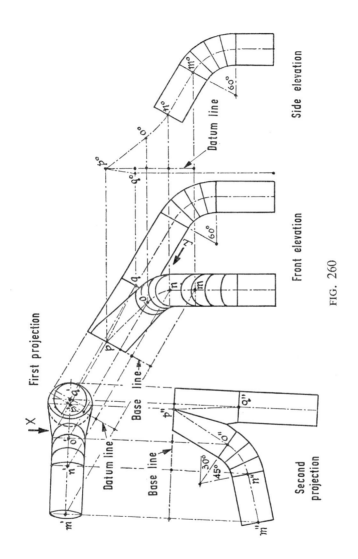

FIG. 260

and q''. It will be noted that point p'' actually occurs on the base line.

Since in the first projection the points m', n', o', p' and q' all lie on the straight line from m' to q', then in the second projection the points m'', n'', o'', p'' and q'' all lie flat in the plane of the paper. This

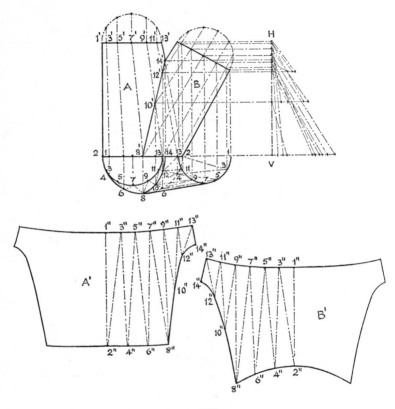

FIG. 261

condition fulfils the purpose of the projection, and the breeches piece with the attached lobster-back bend is now set out in the flat as shown in the Figure. The angle contained by the three segments of the lobster back is 45°, and the angle which the adjoining edge of the breeches piece makes with the base line is 30°. From these conditions the breeches piece is set out for development as in Fig. 261.

THE BREECHES PIECE

Since this branch piece is of the breeches type, the joint line between the limbs is elliptically curved, and as the top edge of the limb A on the left-hand side is parallel with the base of the breeches piece, that limb is a frustum of an oblique cone. The apex of the cone is so far above the top of the limb that development of the pattern by the radial line method would be somewhat troublesome, and the pattern is therefore developed by triangulation. A half plan of the breeches piece is attached to the base line of the elevation.

Preparatory to the development of the patterns, a semicircle is drawn on the top edge of the left-hand limb A, divided into six equal parts, and the points dropped perpendicularly to the edge. Similarly, the semicircle representing the base of the breeches piece in the half plan is divided into six equal parts and the points projected vertically upwards to the base line. Then the corresponding points on the top edge are joined to those on the base line, thus producing the radial lines on the left-hand limb.

Next, to determine the joint line in the plan, the semicircle which represents the top edge in the plan is divided into six equal parts, and the points are joined to the corresponding points on the semicircle which represents the base. The points on the top edge in the plan are then numbered 1, 3, 5, 7, 9, 11, 13, and those on the base are numbered 2, 4, 6, 8, 6, 4, 2. In the elevation, the two radial lines from points 9' and 11' on the top edge cross the joint line at points 10' and 12'. Points 10' and 12' are now projected vertically to obtain points 10 and 12 on the corresponding radial lines in the plan. Thus the curve 8,10,12,14 drawn through these four points in the plan represents the joint line in that view.

To develop the pattern for the left-hand conic limb A, the plan is divided into triangles by the zigzag line 1,2,3,4 . . . 9,10,11,12,13, 14. Note that from point 9, the zigzag line is joined to the points 10, 12 and 14 on the joint line (instead of, as is sometimes done, to points 6, 4, 2 on the base). The procedure for the development of the pattern is straightforward triangulation, and it is assumed that the reader will be able to follow the process from the illustration without difficulty. However, it should be noted that the true distances between points 8 to 10, 10 to 12, and 12 to 14 are obtained by triangulating these plan lengths against their respective vertical heights. The pattern for this limb is shown developed at A'.

The other limb B is *not* a frustum of an oblique cone, and the pattern is therefore developed by the method of triangulation. The inclined top edge is dropped into the plan as an ellipse, and the points thereon, beginning at the outside edge, are numbered in a similar manner to the corresponding points on the limb *A*. Again, as the procedure for the pattern development is fairly straightforward, detailed instructions are not given, though one or two important points should be observed. The true spacings for the top edge are taken from the semicircle in the elevation, and not from the ellipse in the plan. Also, the true spacings between points 8 to 10, 10 to 12, and 12 to 14 are again obtained by triangulating the plan lengths against their respective vertical heights. The pattern for limb *B* is shown developed at *B'*, Fig. 261.

11 Draughtsmanship in Sheet Metal Work

No doubt pattern developing is one of the least expected sources of economy. When it comes to making it pay, when geometry applied to sheet metal work becomes the means of saving money—it is time to take notice. Pattern development is bound up inseparably with sheet metal work. Every article worked out of the flat requires a pattern or template of some sort. One may wink one's eye and say, "Ah, but this template is only a rectangle," or "That blank is merely a circle." Nevertheless, the template or blank has to be measured off or marked out.

Unfortunately, most sheet metal patterns are not so simple, and it is the more intricate ones which cause doubt on the score of time saving. Geometry applied to sheet metal drafting requires real mental effort to probe its perplexities, and this mental effort is a source of trouble where geometrical developments are concerned. How often are air duct systems and arrangements of pipework constructed on lines of simplicity rather than efficiency! Moreover, if we be honest, we must admit that the aim is not so much to simplify the method of construction, as to simplify the geometrical design.

If a system of ductwork is well designed by the draughtsman but contains a few geometrical posers in its construction, and if the draughtsman insists on the work being carried out exactly to drawing on the score of efficiency, what are the chances that the work will cause a stir and possible confusion in the workshop? No craftsman can make something which is beyond his comprehension. No sheet metal worker can produce something which he cannot understand. If he makes the attempt, he will either make a hash of it or alter the job to suit his own convenience. Often enough discussion follows discussion, this or that might be modified, something else might be altered to suit—not always with the object of improving the system. The real difficulty, as a rule, is not the interpretation of practical operations, but the understanding of the geometrical principles involved.

Fig. 262 is a typical example of simplified geometry at the expense of efficiency. The illustration represents a piece of pipework from a fan. The objects to be accomplished are: (1) a transforming piece from a rectangle to a circle, (2) a branch pipe, and (3) a reducing piece to correct the cross-sectional area. The form of construction shown in Fig. 262(a) is perhaps more common than the one shown in Fig. 262(b) because its geometry is simpler and easier to understand.

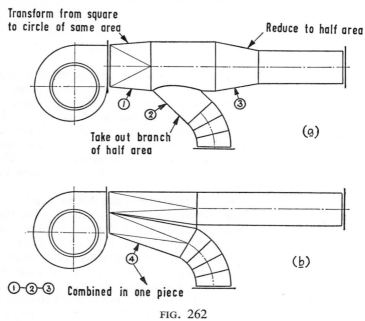

FIG. 262

Nevertheless, the actual work of construction involved in the latter example is much less than that in the former for it combines the objects (1), (2) and (3) in a single piece as shown at (4). Moreover, it is more efficient and more pleasing to the eye. This is only one example of many which might be given of efficiency achieved by the judicious application of geometrical principles. Craftsmen often shun methods which appear complicated, but things only appear complicated when they are imperfectly understood. The deficiency lies not in the method but in the craftsman.

SETTING OUT LARGE OBJECTS

Generally a craftsman cannot efficiently set out a large geometrical problem full size on sheets of metal or on the floor of the depart-

Draughtsmanship in Sheet Metal Work 347

ment. It is sufficient to see him manœuvring with trammels or string in his endeavours to get true lengths for the pattern to know that the method is a time waster. A small conical vessel may reasonably be dealt with directly on the metal itself, but a complex hood of large dimensions will tax all the ingenuity of the craftsman to develop the pattern satisfactorily.

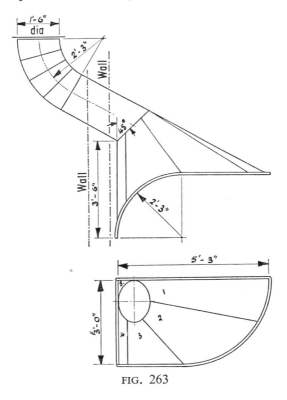

FIG. 263

In dealing with a large hood such as the one shown in Fig. 263 it is inevitable that a pattern must be obtained or arrived at by some means. The old rule-of-thumb method of building it up by guesswork is not to be thought of. To tackle it full size by methods of geometry would be cumbersome. It would be much better, with pencil and paper, to develop the pattern as accurately as possible to some convenient scale, say one-quarter, and then transfer the pattern from the paper to the metal by taking each measurement four times the scale size.

It will be found that patterns developed in this way can be dealt with much quicker and with greater reliability than by attempting to set them out full size. By this method it can also readily be seen how best to apply the pattern to the sheets in order to save material. Very often, by the judicious placing of the patterns, two or more may be cut from one sheet instead of cutting to waste as is so often the case with direct marking out.

Developments to scale in this way may be carried out by the men in the shop if suitably equipped. A better arrangement still is to employ a specially qualified draughtsman who can readily and rapidly deal with the setting out of patterns and dimension them so that the craftsman may easily transfer them to the metal with the highest degree of economy.

It may be argued that this is essentially draughtsman's work. It is. But nearly all draughtsmen are trained on lines of mechanical drawing and machine construction, and very few are qualified to deal with complex surface developments and the preparation of patterns unless specially trained by the sheet metal industry.

A DRAUGHTSMAN'S JOB

The draughtsman should be attached to, and work in, the department concerned, so that he is always on the spot to undertake any measuring up or pattern development immediately it is required. By having one person specialize in this work a great deal of time can be saved by the ready skill and technical efficiency of the expert, while at the same time improved constructive design must result.

Because the draughtsman engaged on this work is not directly productive, it may be thought that the work is an extra charge on the production of the department. But is it? It must be remembered that all work has to be measured up, and patterns have to be developed, by somebody. It is surely safer and quicker for this work to be done by trained and competent hands than by people whose technique is unreliable. An expert practical man may often be lacking in geometrical skill, so that while his powers of execution are excellent in one direction, he is slow and uncertain in the other. One good technician can facilitate the work of a score of practical men. Sometimes days, and even weeks, of time may be saved on a big scheme by the skilful planning and preparation of the work so that the craftsmen may go ahead with confidence instead of spending half their time in doubt and uncertainty on work which really requires the skill of a technician. The essential condition is that the draughtsman must be on the spot as a member of the department, so that his services are readily available.

Draughtsmanship in Sheet Metal Work

DEVELOPMENT OF THE HOOD

Fig. 264 shows the pattern for the hood (Fig. 263) developed to a scale of one twenty-fourth, or $\frac{1}{2}$ in. to 1 ft. Each line in the pattern is dimensioned full size, so that the craftsman may readily mark out each part. The full pattern will require several sheets of metal 6 ft long by 3 ft wide in order to cut out the whole of it. The pattern is therefore divided into five sections with the seams placed so that they occur at convenient positions for working up—this is an important consideration. It will be seen from the diagram that the five sections can be cut from three sheets of metal by carefully placing them as shown.

It must be readily apparent from this diagram that should the attempt be made to mark out the hood full size, several more sheets must inevitably be cut to waste owing to the difficulty of foreseeing the true shape of the pattern and exactly where to begin on the sheets. It will hardly be necessary to mention also the economy of time effected in dealing with this problem to a reduced scale.

No doubt many large firms already employ draughtsmen on the design of sheet metal work, yet do the draughtsmen in question effect this economy by fully planning the work, or do they merely set schemes on paper and leave the real planning to the craftsmen?

DRAWINGS FULLY DETAILED

The short connecting pipe shown in Fig. 265 is presented in ordinary plan and elevation. It will be seen that all necessary measurements are given for determining the relative positions of the two angle frames to be connected up, but no details of the pipe itself are shown. The information necessary for making the two transformers, one at each end, and also for the sections of pipework in the middle, is not complete. It is true that many sheet metal craftsmen who are well acquainted with this type of work would have little or no trouble in making the pipework to suit the conditions required. Nevertheless, in the event of a drawing such as this being supplied for production (and this is by no means uncommon), further detailing would have to be done and additional measurements determined before the work could be set out for development. The efficiency with which this further detailing is done must depend on the skill of the draughtsman or the craftsman and the extent of his knowledge of the geometry of sheet metal work.

The drawing presented in Fig. 266, the same piece of work, shows two auxiliary projections which are necessary to include all

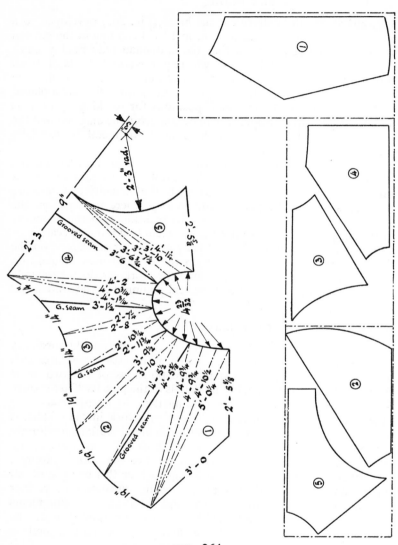

FIG. 264

Draughtsmanship in Sheet Metal Work

the details of the parts which go to make up the piece of pipework between the two given positions. This arrangement is by no means the only construction which could be made to fulfil the given conditions, but the design is one which provides for smooth and easy flow.

To set out the projections and the details of the pipework from the particulars given in Fig. 265, the centreline PQR is first drawn

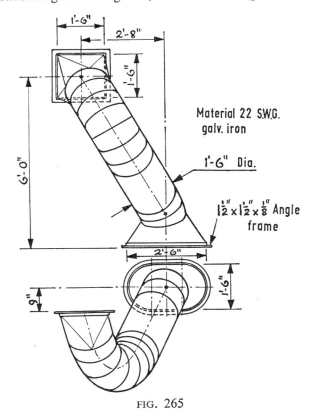

FIG. 265

at 30° to meet the vertical centreline from S in point R. The next step is to obtain the projection in the direction of the arrow M at right angles to PQR, in which $O'R'$ is set off as the base line at right angles to the direction of projection. According to the plan, the base of the square-to-circle transformer will be 9 in. above the point O', which should now be located. The position of the lobster-back bend should now be determined, the first step towards which is the

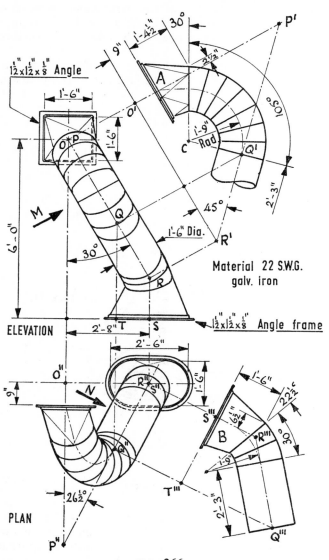

FIG. 266

Draughtsmanship in Sheet Metal Work

setting up of the triangle $O'R'P'$. The angle at R' might be given any value according to circumstances, but in this particular case 45° suits the conditions quite well. Alternatively, if the triangle $O''R''P''$ in the plan has already been decided upon, the distance $O''P''$ should be marked off from O' to P' in the projection, in which case the value of the angle at R' will be determined by this height. However, assuming that the angle of 45° is to remain, then the distance $O''P''$ in the plan must be made to correspond with the height $O'P'$ in the projection. The centre C, for the radius of the bend, is the next point to locate, and this should be placed so that the centreline of the bend touches $R'P'$ at some point Q', while the other end of the centre line points to the centre of the base of the transformer A. A little adjustment may be necessary to fit this bend in so that it forms a smooth connexion with the tallboy transformer.

The projection in the direction of the arrow N may now be drawn at right angles to $R''P''$. First, the point Q'' on $R''P''$ is located by projecting Q' from the first projection back to Q in the elevation, and then dropping Q vertically downwards to Q'' in the plan. Now Q'' represents the centre of the circular cross-section at the end of the lobster-back bend. Therefore in the projection at N, point Q'' may represent the beginning of the straight portion, as will be seen at Q'''. Then $S'''T''''$ is marked off for the base line, and from T'''' the distance $T''''Q'''$ is marked off equal to the height TQ from the elevation. Next, from S'''' the distance $S''''R'''$ is marked off equal to the height SR from the elevation. Then $S''''R'''Q'''$ represents the projected centrelines, and the radius for the two bottom lobster-back segments may be drawn in, together with the projection B of the bottom transformer. The necessary measurements and details of construction may now be drawn in.

PATTERN DEVELOPMENTS TO SCALE

The development of the patterns for the transformers and the lobster-back segment is a task which is almost invariably left to the craftsman, yet both time and material can often be saved by the drafting of patterns to scale. This applies particularly, as already shown, to such examples as hoods and hoppers and many other classes of work of large dimensions. To develop outsize patterns in the workshop demands not only a reliable knowledge of geometry, but also considerable skill and ingenuity in dealing with work to full size dimensions. The setting out of plans and elevations in full on standard size sheets, 6 ft by 3 ft, is often no easy task when these views are larger than the sheets themselves. In this connexion it should be remembered that the metal sheet is the craftsman's drawing board, and to his credit it must be admitted that much skilful work is carried out on it without the aid of precision instruments. Nevertheless, in such circumstances it is almost impossible to anticipate exactly where the pattern should begin in order to

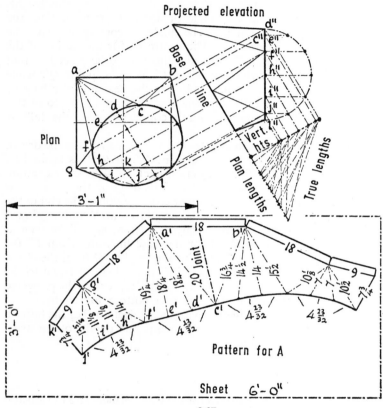

FIG. 267

ensure the most economical layout on the sheets of metal. Consequently, large size work developed in this way is invariably extravagant in the use, or rather waste, of material.

Fig. 267 shows the pattern for the square-to-circle transformer developed by the method of triangulation. This work, done on the drawing board to a suitable scale, with the lines on the pattern dimensioned full size so that they can be readily transferred to the metal sheet, is not only more convenient in the matter of execution

Draughtsmanship in Sheet Metal Work

but should also result in an appreciable saving of time. Coupled with this, when the pattern has been fully developed, the size and position of the sheet from which it is to be cut may be placed round the pattern, as indicated in Figs. 267 and 268, in order to show the best position for cutting.

This principle of setting out patterns to scale is further illustrated in Fig. 268, which shows the development of the transformer at the

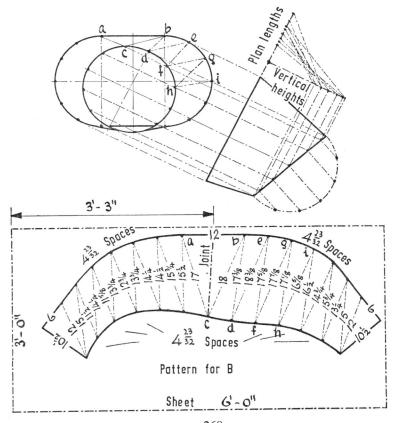

FIG. 268

other end of the duct. Although there is plenty of room on the sheet in both cases, most craftsmen would agree that, in developing these patterns full size direct on the metal sheets, it would be a stroke of luck to get them so well placed for cutting out. When the position of the first line has to be guessed at, there is always a very good chance that the pattern will run off the sheet in the course of development, which would necessitate starting afresh.

It will be noticed that in each of these examples a full sheet is required for the complete pattern. Even though the amount of "scrap" remaining after these patterns have been cut from the sheets is of useful size, the fact remains that two sheets have to be cut into. If, on the other hand these patterns were each made in two pieces with a joint in the middle, as shown at the position of the chain dotted line, both patterns could be cut from one sheet only. This is shown in Fig. 269, which illustrates how material

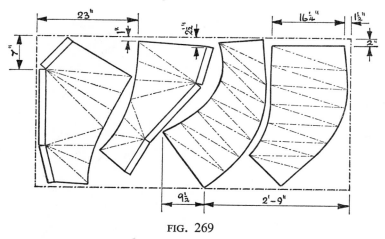

FIG. 269

can be saved by a little more time spent on the drawing. After the patterns have been developed to scale, it is a very simple matter to manœuvre them into the best positions on the sheet.

THE SQUARE-TO-CIRCLE TRANSFORMER

To develop the pattern for the square-to-circle transformer shown in Fig. 267, the surface is divided into triangles as shown in the plan at a,b,c,d,e,f,g,h,i,j,k, which represents one-half of the surface to be developed. In the projected elevation the base line is extended and a vertical height line erected at right angles to it. All the points on $d''l''$ are then projected to the vertical height line as shown in the diagram. For the first line in the pattern, ab is taken direct from the plan and marked off as at $a'b'$ in the most convenient position. Next, the distance bc is taken from the plan and marked off along the base line at right angles to the vertical height. The true length diagonal is taken up to the point level with c'', and from b' in the pattern an arc is drawn through point c'. Now, the distance ac is taken from the plan and marked off along the base line at right angles to the vertical height. The true length diagonal is taken up to the point level with c'', and from a' in the pattern an arc is described cutting the previous arc in point c'.

For the second triangle, the plan length ad is taken and marked off along the base line at right angles to the vertical height. The true

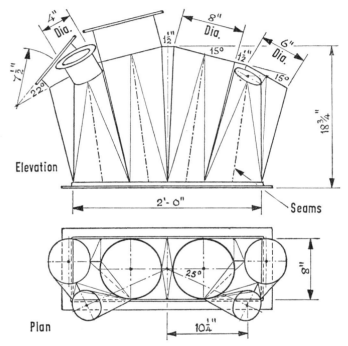

FIG. 270. *A six-branch junction piece*

length diagonal is taken up to the point level with d'', and from point a' in the pattern an arc is drawn through point d'. The next true distance for the pattern, $c'd'$, is taken direct from the semicircle in the projected view, as the spacings around the semicircle are the true distances around the edge of the hole. Therefore one of these spacings is taken, and from point c' in the pattern an arc is described cutting the previous arc in point d'.

This process of triangulation should be carefully followed right round the transformer, as it is not symmetrical about any axis.

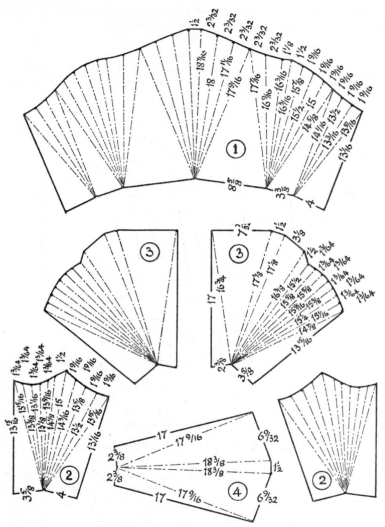

FIG. 271. *Patterns for six-branch junction piece*

Draughtsmanship in Sheet Metal Work 359

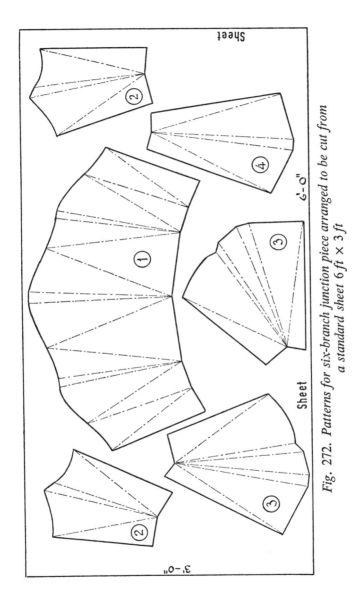

Fig. 272. *Patterns for six-branch junction piece arranged to be cut from a standard sheet 6 ft × 3 ft*

The chief precaution is to see that, when the plan length of the line being dealt with is placed at right angles to the vertical height, the true length diagonal is taken to the correct point on the vertical height line.

The pattern for the other transformer shown in Fig. 268 is obtained in precisely the same way, although the pattern itself is quite different. Directions for its development are not given, as this example is left as an additional exercise in preparing a fully dimensioned pattern.

A SIX-BRANCH JUNCTION PIECE

The six-branch junction piece shown in Fig. 270 may strike the reader as an unnecessarily elaborated example of a branch piece. If so, the author ventures to explain that this was the outcome of an

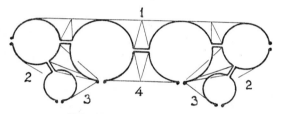

FIG. 273. *Key to patterns* (*Fig. 272*)

actual piece of work designed to suit given conditions in a fume extraction plant. The rectangular flange had to fit against an inlet of similar size on the side of a larger flue containing water sprays. From the rectangular hole it was necessary to branch to six points which were in close proximity to the flue inlet. Several preliminary designs were made, chiefly on the lines of dividing into twos and threes, but the conditions did not lend themselves to good design in dividing and subdividing. The six-branch piece shown in Fig. 270 was therefore designed and the joints, consisting of grooved seams, were arranged to occur on the flat triangular sides as shown in the Figure.

The next important consideration was the practical construction of the branch piece. There were several to be made, all identical, as a number of similar units were to be connected at different points along the spray flue. The work needed skilled craftsmanship. Nevertheless, the units were satisfactorily completed to the above design, and the fume points successfully connected up.

Draughtsmanship in Sheet Metal Work 361

The patterns were developed to scale and dimensioned as shown in Fig. 271. Moreover, the patterns were arranged so that all six could be cut from one standard sheet measuring 6 ft by 3 ft, as shown in Fig. 272.

The diagram in Fig. 273 shows the positions of the six parts when the patterns were formed to shape. In addition to the grooved seams which occur on the flat triangular sides from top to bottom, short joints were necessary along the $1\frac{1}{2}$-in. tops of the smaller triangles.

This example is given to show the advantages of designing for efficiency, developing patterns to scale, and inserting the full size measurements on the patterns so that they can be drawn directly on to the metal sheets ready for cutting.

Detailed directions for developing the patterns are not given as it is assumed that the reader, having progressed to this stage, would have little or no difficulty in arriving at the pattern developments by the method of triangulation.

Index

Auxiliary and double projections, 185
Auxiliary projection, 23

Bowl, elongated segmental, 85
Branch at complex angle, 339
Branch off-centre with main, 208
Breeches pieces, 141
Breeches, three-way, 156
Breeches—
    with branches on same base, 141
    with odd limbs, 149
    with predetermined joint, 145
    with shallow joint line, 147
    with square base, 153

Chute from screw conveyor, 191
Circle to rectangle nozzle, 117
Common central sphere, 97
Complex connexion to circle, 232
Complex hopper, 295
Complex patterns, 254
Cones, two intersecting, 99
Conic sections, 14
Conical cover, sight screen, 182
Conical hopper on conveyor casing, 59
Conical spout on domed vessel, 179
Connexion to angular corner, 134
Connexion to spherical dome, 321
Cross pipes of equal diameters, 75
Curb moulding, 77
Cutting planes, 165
Cylinder to oblique cone, 250
Cylindrical elbow on cylindrical main, 200
Cylindrical pipe on square pyramid, 93

Directrix, 9
Divided feed hopper, 281
Draughtsman's job, 348
Draughtsmanship, 345
Drawings fully detailed, 349
Duct layout with developments, 286

Dust and fluff extraction ducts, 295

Elbow at complex angles, 214
Ellipse, 3
Elongated segmental bowl, 85
Equal tapered tray, 38
Exhaust ducts, 325
Extraction ducts, 302

Feed chute to rotary sieve, 229
Feed hopper, divided, 281
Finial, simple, 82
First angle projection, 20
Flared ventilator head, 264
Flower stand, an ornamental, 84
Focus, 3
Fume ducts, 329
Fume hood and pipe, 125
Fume hoods and extraction ducts, 286

Gusset piece, off centre, 282

Heptagon, 1
Hexagon, 1
Hexagonal oblique prism, 67
Hexagonal right prism, 66
Hexagonal vase, 79
Hopper with extended sides, 300
Hopper with inclined back, 132
Hyperbola, 13

Intersecting cones, 99
Intersection of two square pipes, 194
Isometric projection, 25

Junction piece—
    square to circle, 162
    three-way, 314
    two-way, 160
Junction pieces, 159

Lines in space, 110

Moulding—
  curb, 77
  simple, 75

Non-symmetrical nozzle, 45

Oblique cone, 41
Oblique conic connexion to conic cover, 170
Oblique conic connexion to spherical dome, 176
Oblique conical four-way branch, 56
Oblique conical hood, 48
Oblique conical hopper, 44
Oblique cylinder, 69
Oblique projection, 28
Octagon, 3
Ornamental flower stand, 84
Orthographic projection, 20
Outlet connexion to tank bottom, 73
Overflow chute, 246

Parabola, 9
Pattern development to scale, 353
Pentagon, 1
Pipe connexion to dome, 325
Pipe encircling square column, 242
Points of curves in pattern, 63
Predetermined curves, 269
Projections, 20
Projections—
  on right cone, 190
  on square prism, 187
  on straight line, 185

Rectangle to circle transformer, 128
Rectangular branch on cylindrical body, 91
Rectangular pipe intersecting right cone, 176
Regular polygons, 1
Right conic connexion to conical cover, 167
Right conical connexion to angular surface, 32
Right conical hood, 30
Right conical hopper on inclined pipe, 106
Right conical outlet, 33
Right conical outlet from cylindrical duct, 103

Right cylinder, 68
Right cylindrical chute, 71
Rolled surface transition piece, 256

Segmental bowl, elongated, 85
Segmental lid, 78
Semi-ellipse to circle transformer, 119
Setting out large objects, 346
Simple finial, 82
Simple moulding, 75
Simple off-set pipe, 235
Six-branch junction piece, 357
Spacing error, 114
Spiral chute with centre column, 274
Spiral chute without centre column, 279
Spiral chutes, 254
Spiral finial, 272
Spout on a rectangular can, 36
Square oblique prism, 66
Square pipe elbow on cylindrical main, 198
Square pipe elbow on square main, 196
Square right prism, 65
Square to circle transformer, 113
Stove elbow connexions, 137

Third angle projection, 21
Three-way breeches piece, 156
Three-way conical branch, 54
Three-way junction piece, 313
Transformer connexion, 310
Transition piece, square to rectangle, 254
Triangulation, 109
True lengths by triangulation, 111
Twin branch pipes on main, 211
Twisted square transformer, 111
Two intersecting cones, 99
Two-point right conical nozzle, 101
Two-way junction piece, 160

Ventilator head, 117
Ventilator head, flared, 264
Vertex, 9

Watering can rose, 37

Y-piece inclined at joint, 217
Y-piece with one limb inclined, 220

PJ OBRIEN
8 CEDAR DRIVE
EDENBRIDGE
KENT